Maker Innovations Series

Jump start your path to discovery with the Apress Maker Innovations series! From the basics of electricity and components through to the most advanced options in robotics and Machine Learning, you'll forge a path to building ingenious hardware and controlling it with cutting-edge software. All while gaining new skills and experience with common toolsets you can take to new projects or even into a whole new career.

The Apress Maker Innovations series offers projects-based learning, while keeping theory and best processes front and center. So you get hands-on experience while also learning the terms of the trade and how entrepreneurs, inventors, and engineers think through creating and executing hardware projects. You can learn to design circuits, program AI, create IoT systems for your home or even city, and so much more!

Whether you're a beginning hobbyist or a seasoned entrepreneur working out of your basement or garage, you'll scale up your skillset to become a hardware design and engineering pro. And often using low-cost and open-source software such as the Raspberry Pi, Arduino, PIC microcontroller, and Robot Operating System (ROS). Programmers and software engineers have great opportunities to learn, too, as many projects and control environments are based in popular languages and operating systems, such as Python and Linux.

If you want to build a robot, set up a smart home, tackle assembling a weather-ready meteorology system, or create a brand-new circuit using breadboards and circuit design software, this series has all that and more! Written by creative and seasoned Makers, every book in the series tackles both tested and leading-edge approaches and technologies for bringing your visions and projects to life.

More information about this series at https://link.springer.com/bookseries/17311.

Tiny Machine Learning Quickstart

Machine Learning for Arduino Microcontrollers

Simone Salerno

Apress®

Tiny Machine Learning Quickstart: Machine Learning for Arduino Microcontrollers

Simone Salerno
Trani, Italy

ISBN-13 (pbk): 979-8-8688-1293-4   ISBN-13 (electronic): 979-8-8688-1294-1
https://doi.org/10.1007/979-8-8688-1294-1

Copyright © 2025 by Simone Salerno

This work is subject to copyright. All rights are reserved by the Publisher, whether the whole or part of the material is concerned, specifically the rights of translation, reprinting, reuse of illustrations, recitation, broadcasting, reproduction on microfilms or in any other physical way, and transmission or information storage and retrieval, electronic adaptation, computer software, or by similar or dissimilar methodology now known or hereafter developed.

Trademarked names, logos, and images may appear in this book. Rather than use a trademark symbol with every occurrence of a trademarked name, logo, or image we use the names, logos, and images only in an editorial fashion and to the benefit of the trademark owner, with no intention of infringement of the trademark.

The use in this publication of trade names, trademarks, service marks, and similar terms, even if they are not identified as such, is not to be taken as an expression of opinion as to whether or not they are subject to proprietary rights.

While the advice and information in this book are believed to be true and accurate at the date of publication, neither the authors nor the editors nor the publisher can accept any legal responsibility for any errors or omissions that may be made. The publisher makes no warranty, express or implied, with respect to the material contained herein.

    Managing Director, Apress Media LLC: Welmoed Spahr
    Acquisitions Editor: Miriam Haidara
    Development Editor: James Markham
    Editorial Assistant: Jessica Vakili
    Copy Editor: Kimberly Burton-Weisman

Cover designed by eStudioCalamar

Distributed to the book trade worldwide by Springer Science+Business Media New York, 1 New York Plaza, New York, NY 10004. Phone 1-800-SPRINGER, fax (201) 348-4505, e-mail orders-ny@springer-sbm.com, or visit www.springeronline.com. Apress Media, LLC is a Delaware LLC and the sole member (owner) is Springer Science + Business Media Finance Inc (SSBM Finance Inc). SSBM Finance Inc is a **Delaware** corporation.

For information on translations, please e-mail booktranslations@springernature.com; for reprint, paperback, or audio rights, please e-mail bookpermissions@springernature.com.

Apress titles may be purchased in bulk for academic, corporate, or promotional use. eBook versions and licenses are also available for most titles. For more information, reference our Print and eBook Bulk Sales web page at http://www.apress.com/bulk-sales.

Any source code or other supplementary material referenced by the author in this book is available to readers on GitHub. For more detailed information, please visit https://www.apress.com/gp/services/source-code.

If disposing of this product, please recycle the paper

Table of Contents

About the Author xiii

About the Technical Reviewers xv

Preface xvii

Chapter 1: Tiny Machine Learning 1
   What Is Machine Learning? 2
   What Is Tiny Machine Learning? 3
      What Can Be Considered "Tiny"? 5
      Why Machine Learning on Microcontrollers? 7
      What Can Tiny Machine Learning Do? 10
      What Can't Tiny Machine Learning Do? 12
   Machine Learning Glossary 13
      Supervised vs. Unsupervised 14
      Ground Truth 16
      Regression vs. Classification 16
      Metrics 21
      Metrics for Binary Classification 21
      Overfitting and Underfitting 29
      Training vs. Validation vs. Test 31
      Feature Engineering 33
      Types of Data 34
   Summary 40

TABLE OF CONTENTS

Chapter 2: Tabular Data Classification 41

Required Hardware 43
Required Software 44
    Create A Python Virtual Environment..45
    tinyml4all Arduino Library 46
Capture Data..47
    1. Manually Copy Data from Serial Monitor...49
    2. Read Serial Output from Python 53
    3. Save to SD Card..56
Load and Inspect Data 60
    One File per Class..61
    One File for All Classes 62
    Manipulate Table 63
    Plot Data 66
Feature Engineering..71
Feature Scaling 72
    How to Identify Outliers 74
Feature Selection 78
    Sequential Feature Selection 79
    Score-based Selection 80
    Recursive Feature Elimination (RFE) 80
Classification Models..82
    Decision Tree 84
    Random Forest 86
    Extreme Gradient Boosting...87
    Logistic Regression 88
    Support Vector Machines..91

| Classification Chain | 96 |

Deploy to Arduino 98

    How to Use in Arduino Sketch 98

    Warnings 101

Summary 101

Chapter 3: Tabular Data Regression 103

Required Hardware 105

Capture Data 106

Load and Inspect Data 109

Plot Regression Data 110

    One Input 111

    Many Inputs, Many Scatters 112

Feature Engineering 114

    Monotonic Functional Mappings 114

    Polynomial Input Combinations 117

Regression Models 118

    Ordinary Least Squares 119

    Ridge and Lasso 119

    Decision Tree and Random Forest 120

Regression Chain 122

Deployment 124

    How to Deployment Use in Arduino Sketch 124

Summary 127

Chapter 4: Time Series Classification Using Edge Impulse 129

Required Hardware 131

Required Software 132

TABLE OF CONTENTS

Capture Data 133
    Load And Inspect the Data 137
    Plot Time Series Data 138
Feature Engineering 139
    Time-Domain Features 142
    Frequency Domain Features 143
Edge Impulse for Continuous Motion 145
    Edge Impulse Workflow 145
    CSV Layout Configuration Wizard 149
    Upload Files 153
    Train/Test Split 154
    Impulse Design 155
    Fully Connected Neural Networks 162
Testing 164
Deployment 164
    Edge Impulse Shortcomings 169
Summary 170

Chapter 5: Time Series Classification Using Python 171

Hardware Requirements 172
Capture Data 173
Data Labeling 177
Feature Engineering 179
    Statistical Moments 180
    Autocorrelation 181
    Shape Metrics 181
    Windowing 182
    One vs. Rest 183

Episodic Time Series Classification Chain	183
Deploy to Arduino	186
Summary	189

Chapter 6: Audio Wake Word Detection with Edge Impulse 191

Hardware Requirements	193
Software Requirements	194
Capture Data	194
Audio Data Format	196
Arduino Sketch	196
Python Code	199
Third-Party Datasets	201
Synthetic Wake Word Generation	201
Load and Inspect Data	206
Edge Impulse Data Acquisition	209
Feature Engineering	210
Mel Spectrogram	211
Mel-Frequency Cepstral Coefficients	214
Audio Classification with Edge Impulse	215
Convolutional Neural Networks	216
Testing	220
Deployment	222
Continuous Classification	222
Summary	225

Chapter 7: Object Detection with Edge Impulse 227

| Hardware Requirements | 228 |
| Software Requirements | 229 |

ix

TABLE OF CONTENTS

Capture Data 229
    Arduino Sketch to Collect Images..230
    Python Code to Read Images..232
Edge Impulse Data Acquisition..235
Feature Engineering...238
    Impulse Design...239
Testing..243
Deployment...245
    Visual Debugging..248
Summary..252

Chapter 8: TensorFlow from Scratch 255

Required Hardware 256
Required Software 256
Neural Network Structure 258
    Forward Pass..260
    Backward Pass..260
Multilayer Perceptron..261
    How to Train a Multilayer Perceptron 265
    How to Deploy a Multilayer Perceptron 270
Deep Learning...273
    Deep Learning Disadvantages in the TinyML Context 273
Recurrent Neural Networks and Long Short-Term Memory............................274
    How to Train a Recurring LSTM Neural Network 276
    How to Deploy an RNN 279
1D Convolutional Neural Networks 282
    How to Train a 1D CNN on Continuous Motion...284
    How to Deploy a 1D CNN 286

2D Convolutional Neural Networks 287
    Downsampling and Stride 288
    How to Train a 2D CNN 291
Summary 297

Appendix A: More Feature Engineering Operators 299

Appendix B: References 313

Index 317

About the Author

Simone Salerno has been tinkering with microcontrollers for nearly 10 years and is committed to bringing his knowledge of software engineering to the world of Arduino programming. With the advent of Tensorflow for Microcontrollers he began developing leaner, faster alternatives to neural networks for microcontrollers and started porting many traditional ML algorithms such as Decision Tree, Random Forest, and Logistic Regression from Python to self-contained, hardware-independent C++, ready to be deployed to any microcontroller. Today, he continues to focus on the development of TinyML tools and tutorials with his low-code libraries and no-code online platforms like Edge Impulse.

About the Technical Reviewers

Farzin Asadi received his B.Sc. in Electronics Engineering, his M.Sc. in Control Engineering, and his Ph.D. in Mechatronics Engineering. Currently, he is with the Department of Computer Engineering at the OSTIM Technical University, Ankara, Türkiye. Dr. Asadi has published over 40 international papers and 30 books. His research interests include switching converters, control theory, robust control of power electronics converters, and robotics.

Marcelo Rovai is an educator and professional in engineering and technology, holding the title of Professor Honoris Causa from the Federal University of Itajubá, Brazil. His educational background includes an Engineering degree from UNIFEI and an advanced specialization from the Polytechnic School of São Paulo University (POLI/USP). Further enhancing his expertise, he earned an MBA from IBMEC (INSPER) and a Master's in Data Science from the Universidad del Desarrollo (UDD) in Chile.

ABOUT THE TECHNICAL REVIEWERS

With a career spanning several high-profile technology companies such as AVIBRAS Airspace, AT&T, NCR, and IGT, where he served as Vice President for Latin America, he brings industry experience to his academic endeavors. He is a prolific writer on electronics-related topics and shares his knowledge through open platforms like Hackster.io.

In addition to his professional pursuits, he is dedicated to educational outreach, serving as a volunteer professor at the IESTI (UNIFEI) and engaging with the TinyML4D group and the EDGE AIP – the Academia-Industry Partnership of EDGEAI Foundation as a Co-Chair, promoting EdgeAI education in developing countries. His work underscores a commitment to leveraging technology for societal advancement.

Preface

In the ever-evolving landscape of technology, we find ourselves at a fascinating intersection where the worlds of machine learning and embedded systems converge. This convergence has given birth to Tiny Machine Learning (often shortened as TinyML®, trademark of the EdgeAI foundation)—a field that brings the power of artificial intelligence to the smallest and most resource-constrained computing devices. Rather than requiring expensive cloud infrastructure or powerful computers, Tiny ML allows sophisticated algorithms to run directly on microcontrollers, enabling a new generation of intelligent devices that can make decisions locally, with minimal power consumption and without Internet connectivity.

For Arduino enthusiasts, TinyML represents both an exciting opportunity and a significant challenge. The Arduino platform has long been beloved for its accessibility, allowing individuals with minimal programming experience to create interactive electronic projects quickly. Now, with TinyML, this accessible platform gains the ability to implement sophisticated machine learning capabilities that were previously beyond reach.

However, the journey into TinyML can be daunting, especially for those who may not have a background in data science or machine learning. Many tutorials and resources assume prior knowledge of machine learning principles or require extensive mathematical understanding, creating a significant barrier to entry. Furthermore, the constraints of microcontroller environments add another layer of complexity. Unlike cloud-based machine learning systems with virtually unlimited resources, TinyML development requires careful optimization

PREFACE

to fit within tight memory constraints, limited processing power, and strict energy budgets. Converting and optimizing machine learning models for these environments demands specialized knowledge that bridges the gap between traditional machine learning and embedded systems programming.

This is precisely why this book exists. *Tiny Machine Learning Quickstart* takes a fundamentally different approach to teaching TinyML. Rather than digging too much into the theoretical and mathematical foundations of data science, we start with working solutions and practical code that you can implement immediately from the very first chapters. While understanding the underlying principles of machine learning is undoubtedly valuable, many Arduino enthusiasts simply want to add intelligent features to their projects without becoming machine learning experts. This book respects that desire by providing ready-to-use code snippets, complete workflows, and practical techniques that deliver results first, with theory introduced only where necessary to support practical understanding.

About This Book

Throughout the pages of this book, you'll find a collection of carefully crafted, copy-paste solutions that address common TinyML challenges. Need to build a gesture recognition system? There's a snippet for that. Want to implement keyword spotting to respond to voice commands? You'll find working code ready to be adapted to your project. Interested in object detection from camera images? We've got you covered with implementations you can use right away. As you journey through these chapters, you'll develop hands-on skills that enable you to:

- Implement machine learning models on Arduino boards without needing to understand the mathematical intricacies of each algorithm.

- Follow straightforward workflows that take you from idea to functioning prototype without getting lost in theoretical detours.

- Adapt proven code patterns to suit your specific project requirements, building upon solid foundations rather than starting from scratch.

- Optimize your TinyML implementations for performance, energy efficiency, and reliability using practical techniques rather than abstract concepts.

- Troubleshoot common issues with tested solutions that address the real-world challenges of deploying machine learning on constrained devices.

The book deliberately avoids the trap of trying to transform you into a machine learning researcher. Instead, it treats TinyML as another powerful tool in your Arduino toolkit—one that happens to bring intelligence to your creations. Just as you don't need to understand semiconductor physics to use an Arduino, you don't need to grasp every nuance of neural network design to implement effective TinyML solutions.

Whether you're a hobbyist looking to add intelligence to your weekend projects, an educator seeking accessible ways to introduce students to AI concepts, or a professional hoping to prototype smart devices quickly, this book offers a pragmatic path forward. In a field often characterized by complexity and theoretical depth, "Tiny Machine Learning Quickstart" stands as an invitation to simply begin building—and to learn through the joy of creation rather than the labor of study.

So power up your Arduino, prepare your development environment, and get ready to embark on a hands-on journey into the world of tiny machine learning. The future of intelligent embedded systems awaits, and it's more accessible than you might have imagined.

PREFACE

What You Need to Know Already

This book is geared towards Arduino programmers, so you should already be familiar with Arduino programming, sketch uploading and debugging. For the sake of projects in this book, even a beginner-level experience would be sufficient.

The machine learning part is coded in **Python**: even if most examples are meant to be copy-pasted and run as-is, at least a minor knowledge of the language, how to install packages and how to run a script will greatly help.

Dedication

To my lovely wife Alessia, my heartfilling daughter Adelia and my whole family. They made me the happy person I am today.

# CHAPTER 1

Tiny Machine Learning

Your journey begins! This chapter explores the basics of machine learning and what *tiny* machine learning is all about. You'll learn how it is changing many industries by allowing AI models to run on tiny devices like microcontrollers, sensors, and other Internet of Things (IoT) gadgets. We'll dive into the challenges and opportunities that come with this new frontier technology and discuss how it impacts industries and individuals. The second part of this chapter breaks down the key concepts and terminology you need to know to understand machine learning. It starts with the basics and covers essential terms like supervised and unsupervised learning, classification and regression, and how to measure a model's performance.

Don't worry if you're new to this topic—the chapter takes it one step at a time, covering the fundamentals, and by the end of this chapter, you'll have a solid grasp of the basics. If you're already familiar with machine learning, you can skip ahead, but make sure to at least glance over the headings to ensure you're not missing any crucial concepts.

Understanding this chapter is mandatory for grasping the rest of the book, so take your time and absorb all the information.

CHAPTER 1 TINY MACHINE LEARNING

What Is Machine Learning?

Machine learning is a field of study that enables computers to learn without being explicitly programmed.

—Artur Samuel, 1959

This definition (one of the many available on the subject—and maybe the one that most directly conveys the point) contrasts with the traditional approach to computer programming, where you provide instructions to the computer, which then executes them (see Figure 1-1).

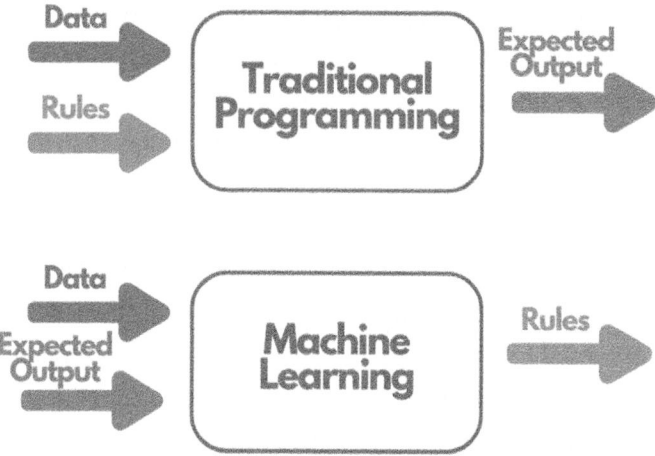

Figure 1-1. *Traditional programming vs. machine learning*

In many cases, it is challenging to come up with exact instructions to achieve a specific goal in software development. For instance, consider a program that detects spam emails. If you were to write such a detector by hand, you would likely look for words that are commonly associated with spam, such as *free, amazing,* or *you are the winner*. However, this approach would require enumerating all possible spammy words and developing a formula for a spam score. This task can be time-consuming and may not be feasible with a satisfactory degree of accuracy.

Machine learning is a relevant approach when it is unclear how to classify information or extract new knowledge from existing data. It systematically analyzes data to identify patterns that may not be immediately apparent to the human eye. After a model has been trained on its task, it can be used to make *inferences* (predict a result based on the input data).

There are hundreds of different ways a computer can detect and extract patterns from data; since this is not a book about machine learning theory, I only briefly touch a few of them when needed. Besides detecting undercover patterns in data once, machine learning can be run periodically without human intervention, ensuring that its decision-making capabilities remain up to date, which is a huge advantage.

Historically, the power of machine learning algorithms has increased over time, largely due to the growth in the size and complexity of the models. This has led to an increasing need for resources, such as RAM and CPU/GPUs, which can be challenging when working with embedded or microcontroller-based systems. However, it does not mean machine learning cannot be run on tiny hardware.

What Is Tiny Machine Learning?

If you're reading this book instead of a book about *traditional* machine learning, it means you're working in the embedded world. By *embedded*, I mean hardware that focuses on a single task (as opposed to a desktop PC that is meant to run a wide range of software for many different purposes) and can interact with the surrounding environment through sensors and actuators. Due to size, efficiency, power, and cost constraints, embedded hardware is characterized by limited resources (CPU, program space, and RAM). Even though the word *embedded* can cover a wide range of devices, in this book it refers to microcontroller hardware (a.k.a. MCUs).

That said, *tiny machine learning*, also known as TinyML (trademark of the EDGE AI FOUNDATION), is a specialized area of machine learning that focuses on optimizing models to run on our embedded devices. This optimization can take two forms.

- Creating custom models that are designed to be as small as possible
- Adapting existing models through compression to fit within hardware constraints

As you might expect, not all models can be compressed to run on such limited resources, but you will be surprised by the many possibilities still available to you. These options are examined in this book.

The fundamental enabling factor for TinyML is an asymmetry in the machine learning process: the resources required to execute a model (give input data and ask for output—inference) are typically orders of magnitude less than the resources required to train a model (give input data and ask to discover patterns—training). We can exploit this asymmetry by training our models on resource-heavy hardware, such as a desktop PC, and then converting them into a lightweight format that fits the constraints of our embedded hardware.

While there exists the possibility to make the training work on devices with limited resources under specific constraints (on-device learning and federated learning), this book focuses on the asymmetric workflow since it gives us a lot more freedom in the choice of the models, allows for larger training data sizes and simplifies the overall development by leveraging well-tested software and tools (on-device learning and federated learning are still in their infancy).

The first part of each chapter is executed on your powerful PC with plenty of resources. You experiment, make changes, and iterate until you get results that satisfy the quality requirements. The second part of each chapter shows you how to generate C++ code that runs on the microcontroller and only performs inference.

CHAPTER 1 TINY MACHINE LEARNING

What Can Be Considered "Tiny"?

The definition of TinyML has undergone significant changes over the years. My initial approach to TinyML was a blog post on the TensorFlow website titled "How to Get Started with Machine Learning on Arduino." The post showcased several machine learning projects running on an Arduino Nano BLE Sense, which features an ARM Cortex-M4 CPU, 1 MB of flash memory, and 256 KB of RAM. Since then, two notable developments have occurred.

TinyML has become synonymous with TensorFlow for Microcontrollers, a specific version of TensorFlow designed for embedded hardware. This is largely due to advertising efforts by the TensorFlow team and vendors that have invested in the technology.

The maker community has successfully run TensorFlow on a range of microcontrollers and more powerful hardware, including the Raspberry Pi and NVIDIA Jetson Nano. This has led to the coining of a new, more encompassing term: EdgeAI.

In this sense, TinyML is a subset of the more general field of EdgeAI, which is a subset of AI. The remainder of this book doesn't cover EdgeAI (AI on portable or mobile devices with a CPU at gigahertz frequency), but only TinyML (AI on bare metal hardware with a CPU in the megahertz range).

An alternative perspective for defining the boundaries of what constitutes "tiny" in TinyML is power consumption. The EDGE AI FOUNDATION is a community focused on ultra-low power machine learning at the edge. However, this term is not precisely defined and can be interpreted in different ways. Typically, it refers to power consumption in the milliwatt range. With this definition, the Raspberry Pi Zero (1 GHz CPU and 512 MB RAM) would fall under the *tiny* umbrella considering its average power consumption of 500–700 mW; but it would be at the high end of the power spectrum, consuming ten times the energy of a typical microcontroller. Figure 1-2 shows examples of what is considered "tiny" or not under this definition.

Figure 1-2. Examples of "tiny" and "not tiny" hardware

For this reason, let's establish a set of soft boundaries for what is and is not considered tiny in the context of this book.

Note This book considers valid TinyML hardware microcontrollers with at most 8 MB of program memory and 4 MB of RAM.

Table 1-1 lists a few examples of TinyML and EdgeAI devices with their hardware specs.

Table 1-1. Hardware Specs for TinyML and EdgeAI Devices

Device	Hardware Specs	TinyML or EdgeAI?
Arduino Nano and Nicla family	ARM Cortex M CPU at 64-480 MHz 256 KB — 1 MB RAM	TinyML
ESP32S3	Xtensa dual-core CPU at 240 MHz 512 KB RAM + external 1 MB RAM (optional)	TinyML
Teensy 4.1	ARM Cortex M7 CPU at 912 MHz 1MB RAM	TinyML
Raspberry Pi 5	Broadcom BCM2712 quad-core CPU at 2.4 GHz Up to 16 GB RAM	EdgeAI
NVIDIA Jetson Nano	ARM Cortex A57 CPU at 921 MHz 4 GB RAM	Edge AI

Does this mean you won't be able to run the code listed in this book on more powerful targets? No, it doesn't. In most cases, you can run the generated code on any hardware that supports C++. However, the primary focus of this book is to showcase the optimizations and trade-offs that differentiate tiny machine learning from traditional machine learning. If your hardware supports more powerful models and algorithms, you need a compelling reason to prefer less capable ones.

Why Machine Learning on Microcontrollers?

Most machine learning applications today are designed to run on powerful hardware, often in cloud computing environments with (virtually) unlimited resources. So, why consider running machine learning on tiny, resource-constrained microcontrollers (MCUs)? Under what circumstances do microcontrollers outperform desktop PCs or cloud computing?

Bandwidth

Connectivity is a fundamental aspect of our digital lives. We are constantly connected, and most of our data is transmitted over networks. However, this is not always the case for embedded hardware. Many microcontrollers lack radio capabilities and rely solely on their internal components to perform tasks. If they want to perform machine learning, they must do so locally.

Even if they can connect to more powerful devices, it does not mean they can transmit all the data they collect. Transmitting data can be costly, both in terms of money and power consumption. Networking is often the most power-hungry component of embedded hardware, and your device may be operating on a limited energy budget (e.g., batteries). It is more practical to perform machine learning inference locally and transmit only the results to a remote location (e.g., for monitoring purposes).

Latency

Transmitting data also incurs time costs. Embedded hardware often requires real-time processing (e.g., self-driving cars or industrial control systems). If your device relies on a remote server for instructions, it may fail due to network latency and create a regulatory disaster (e.g., a car accident or machinery malfunction).

Local machine learning may be slower than transmitting data to a remote server and waiting for a response. However, it is predictable. The time inference takes to complete is typically consistent across runs. You can benchmark this time and factor it into your system's development. Yet, network time is unreliable and often fluctuates based on network conditions outside your control.

Privacy

Not all data is sensitive, and users may not be concerned about transmitting it over the internet for processing. However, sensitive data should not leave the device that collected it. Audio and video data, for example, may be considered private, and many individuals are uncomfortable knowing that their data is exposed to third-party companies and susceptible to leaks or breaches.

In some industries, such as security, health, or childcare, strict privacy regulations must be followed. In these cases, it is preferable, if not mandatory, to keep data local and analyze it on the same device that collected it.

Ubiquitous Computing

The rise of the Internet of Things (IoT) has led to an explosion in the number of smart electronic devices surrounding us. We are surrounded by thousands of objects with microcontrollers, such as light bulbs, ovens, cameras, and thermostats. Due to size and power consumption constraints, these objects cannot afford large or power-hungry motherboards. They must be affordable to keep the end price as low as possible. They must be self-contained because they are deployed in unpredictable environments. TinyML enables each object to implement intelligence into its operating mode, ranging from simple environment understanding (e.g., detecting when a room is empty) to more complex tasks like speech recognition.

Low Power

Microcontrollers are characterized by their low power consumption. To put this into perspective, the following are common machine learning hardware's typical power consumption levels.

CHAPTER 1   TINY MACHINE LEARNING

- Microcontroller: 0.1–0.5 W
- Raspberry Pi: 2–5 W
- Jetson Nano: 5–10 W
- Desktop PC with GPU: 10–50 W

If your deployment has a low power budget (e.g., you are off-grid and rely on batteries or solar energy), you cannot afford to supply watts of power continuously. You are forced to choose the least power-demanding solution. This solution is usually represented by microcontrollers, which often provide ultra-low power modes that reduce their consumption even lower to microwatts, greatly extending the battery life. If your project requires lightweight, sporadic machine learning that runs at regular intervals or only when new data is available, you can avoid wasting power by keeping an oversized piece of hardware idle most of the time.

What Can Tiny Machine Learning Do?

Despite its small size, machine learning on resource-constrained devices can still achieve impressive results. Don't be hasty in thinking that tiny is synonymous with useless. Many industries and verticals have already been impacted by TinyML, and it has proven to work reliably and satisfactorily in widespread real-world applications.

Human Activity Recognition

Smartwatches have made this use case popular in recent years. Thanks to the onboard accelerometer, they can detect the activity you are performing (running, walking, swimming, etc.). They've also been tuned for the elder assistance niche, where they can detect elders' falls and alert caregivers accordingly. This kind of device must obey a large number of constraints (consumes little power to save battery and prevent overheating, and runs fully locally because connectivity is not guaranteed), yet it works so well that it has become a mainstream gadget for people who do sports and fitness. Chapters 4 and 5 approach the task of time series classification using distinct tools and techniques.

Keyword Spotting

The phrases "Hey Alexa" and "Hey Google" have become familiar in recent years. Devices that respond to our voice commands enter our homes and smartphones. To provide a high level of speech recognition accuracy, these devices require an Internet connection to perform the word recognition in the cloud. That said, it would pose serious privacy concerns if these devices were streaming *everything to remote servers*. They instead look locally for the magic activation words, and only when those words are found do they start streaming. Chapter 6 leverages the Edge Impulse platform to train a model that recognizes the magic words "Hey Arduino" fully locally.

Image Classification and Object Detection

Image recognition is an area where neural networks and deep learning have shown impressive results compared to humans. These models have been used in various industrial settings, such as computer-vision-guided machinery in industrial plants to identify and sort products or self-driving robots and cars.

Computer vision models tend to have a high degree of complexity, given the vast number of subjects that an image can contain. Nevertheless, some powerful yet lightweight architectures have been proposed in recent years that are suitable for the TinyML environment. Chapter 7 leverages the Edge Impulse platform to perform object detection (which objects are present in the image? Where are they located?). If a couple of years ago these models would have required a few seconds to run on commodity microcontroller hardware, nowadays you can perform object detection on 10 USD hardware at an astonishing speed of up to 20 frames per second.

Predictive Maintenance

One key application of TinyML in industrial settings is in the monitoring of equipment, such as pumps, motors, and compressors. TinyML-powered sensors can be deployed to monitor equipment performance to detect early signs of wear and tear and predict when maintenance is required. This enables industrial operators to schedule maintenance during planned downtime, reducing the risk of unexpected equipment failures and minimizing the impact on production.

What Can't Tiny Machine Learning Do?

Not all machine learning tasks are suitable for tiny computers. Many recent advancements in the field often struggle to run on consumer-level desktop hardware, let alone tiny machines.

Large Language Models

In November 2022, ChatGPT revolutionized the artificial intelligence world by demonstrating its ability to understand human instructions and respond in a human-like style. Since then, large language models (LLMs) have gained significant attention, and their development has accelerated. These models are characterized by their massive size, often with billions (or even trillions) of parameters and requiring billions of mathematical operations. Unfortunately, such models are unlikely to fit on 2 MB of RAM now and for the years to come.

Image Generation Models

Similar to language models, image generative models, such as Stable Diffusion, are renowned for their massive resource requirements. These models are designed to run on graphics processing units (GPUs) with tens of gigabytes of RAM. Even simpler models can take seconds or minutes to run on everyday laptops, making it impractical to run them on microcontrollers in the near future.

Point Cloud and LIDAR

In the context of Arduino projects, you may want to program an autonomous vehicle driven by a LIDAR sensor, as many autonomous vacuum cleaners do. Although this application area may seem suitable for embedded hardware, reconstructing a 2D/3D scene from the LIDAR point cloud is a math-intensive operation that requires a powerful processing unit. While the CPU does not need to be massive, you still need a modest-sized one—like a Raspberry Pi.

Now, let's focus on the fundamental concepts of machine learning. This is essential because these concepts are explored throughout the rest of the book.

The section that follows may appear like a list of definitions, and you may be inclined to skip it. However, do not do so. These entries are not simply definitions but rather a concise explanation of key concepts that are crucial to understanding the subsequent chapters. These concepts are presented in a clear and accessible manner, making them easy to grasp, even if you are new to the subject.

Machine Learning Glossary

Before moving on to practical code examples in the next chapters, you must become familiar with a few terms and concepts that are the cornerstones of machine learning. Without this prior knowledge, you won't be able to follow the theory and practice that comes in the next chapters. So, take some time to assimilate the following paragraphs as best you can.

Remember that most TinyML comes *after* the traditional machine learning workflow: you first develop a model and then optimize it to run on embedded hardware. If you don't understand the basics of machine learning, you just can't do tiny machine learning.

Note The following paragraphs are not meant to perfectly match the scientific definitions in a math-oriented textbook. They're intended to convey the point with an accessible language instead.

Supervised vs. Unsupervised

Let's revisit the definition of machine learning I introduced at the beginning of this chapter.

> *Machine learning is the field of study that enables computers to learn without being explicitly programmed.*

How does a computer accomplish a task without being programmed? In machine learning, the computer learns from the data it is given. Data can take many forms, such as sensor data, images, audio, or raw numbers. The underlying assumption of the machine learning approach is that this data contains patterns that may or may not be obvious. However, these patterns must exist for any software to be able to learn something useful.

If the data is organized into existing categories or it is known *a priori* what output is expected given an input, it is *supervised learning*. In supervised learning, you provide the model with labeled data, saying, "When the input looks like this, the output should be that." After the training phase, your objective is for the generated model to produce the correct output on new, unseen data. Each model has its unique way of learning this relationship, and each algorithm has varying modeling capabilities. No single model works best for every project, so you need to try a few before finding the optimal one for the task.

CHAPTER 1 TINY MACHINE LEARNING

On the other hand, *unsupervised learning* does not involve outputs. You only have input data and are wondering if it contains naturally occurring patterns. These patterns may not exist at all, or they may be blurred and fuzzy. Your objective in unsupervised learning is twofold: first, to determine whether any pattern exists, and second, to assess how well it fits the data. Examples of unsupervised learning include clustering (grouping similar data together), dimensionality reduction (compressing data without losing too much information), and association rule mining (identifying patterns between events).

Figure 1-3 helps clarify the distinction.

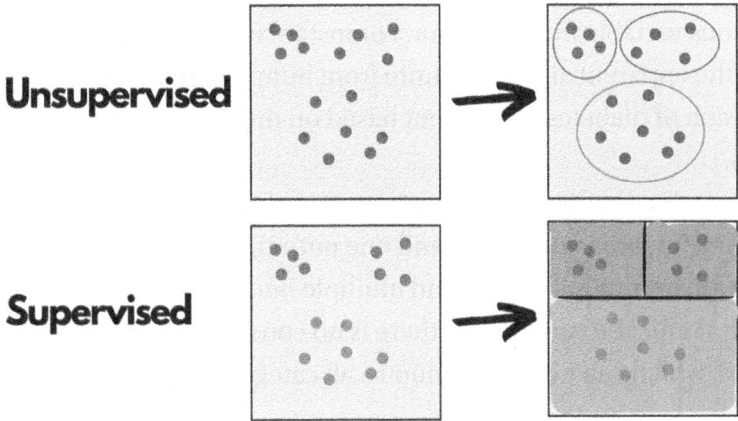

Figure 1-3. *Supervised vs. unsupervised learning*

Note This book only covers supervised learning. It is the most exploitable for real-world projects and lends itself better to tiny machine learning applications.

Ground Truth

In supervised learning, ground truth is the correct output of your input data. The model uses it to learn what the result should look like when presented with a given input. After the model has learned its parameters, the ground truth values are compared with the predicted values to assess the learning performance.

Regression vs. Classification

Regression, a form of supervised learning, is the task of predicting a continuous variable from an input. For instance, you might use regression to infer the apparent air temperature from humidity readings or predict the progression of diabetes in a patient based on their age, weight, and blood analysis.

You can have various combinations of inputs and outputs in regression, including one input and one output, multiple inputs and one output, or multiple inputs and multiple outputs. While the output variable(s) must be continuous, there is no constraint on the input variables, which can be all continuous, all categorical (discrete), or a mix of both.

Classification, part of supervised learning, identifies to which class or classes an input belongs. This task has several important specifications.

- An input can belong to none, one, or multiple classes.
- The list of classes is completely known, meaning a model cannot detect a class it has never seen before.

The task is a *one-label classification* if all input samples belong to exactly one class. If samples can belong to one or more classes, it is called *multi-label classification*. To simplify development, we assume that a sample belongs to at least one class. If it does not, you can discard the sample or create an "unknown" class that groups all samples without a label.

CHAPTER 1   TINY MACHINE LEARNING

An example of classification is determining whether an image depicts a dog or a cat or whether an audio sample contains the words "Hey Google" (with "contains" and "does not contain" labels).

See Figure 1-4 for a visual explanation of the difference.

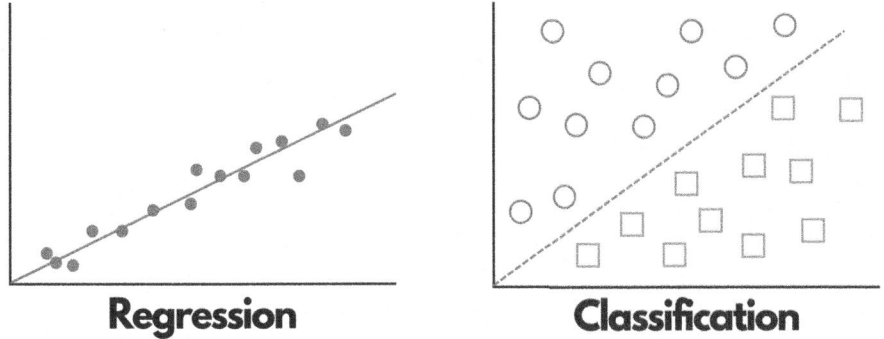

Figure 1-4. Regression vs. classification

Note   This book focuses on regression and one-label classification.

Binary vs. Multiclass

Binary classification is a type of classification problem where you have only two classes. These classes can represent anything, such as cat vs. dog, spam vs. *ham* (not spam), or healthy vs. sick. In binary classification, you assign a numeric value of 0 or 1 to each class. It is a common convention to assign 1 to the positive class and 0 to the negative class. But, this assignment is arbitrary in cases where the classes do not have a natural positive or negative interpretation, such as dog vs. cat.

Multiclass classification, on the other hand, is a type of classification problem where you have more than two classes. There is no theoretical limit to the number of classes, and some classification problems, such as image recognition, can involve thousands of classes. However, the more classes you have, the more complex the classification problem may become.

Some classification models are designed to work with multiple classes, while others are designed to work with only two classes. However, binary-only classification models can be extended to work with multiple classes using one of two strategies: *one vs. all* or *one vs. one*.

One vs. All

The *one vs. all* strategy involves training a separate binary classifier for each class (see Figure 1-5). Each classifier is trained to distinguish between the class of interest and all other classes. The final prediction is made by selecting the class with the highest confidence score.

For example, in a three-class problem (A, B, C), the one vs. all strategy would involve training three binary classifiers.

- Classifier 1: A vs. (B or C)
- Classifier 2: B vs. (A or C)
- Classifier 3: C vs. (A or B)

This strategy only works for binary classifiers that produce a score or probability for their predictions (not all models do). For each new input sample to be classified, all the classifiers make a prediction, and the positive prediction with the highest score is chosen. If no classifier makes a positive prediction or the score is below a confidence threshold, the sample should be marked as *unknown* or *unclassifiable*.

CHAPTER 1 TINY MACHINE LEARNING

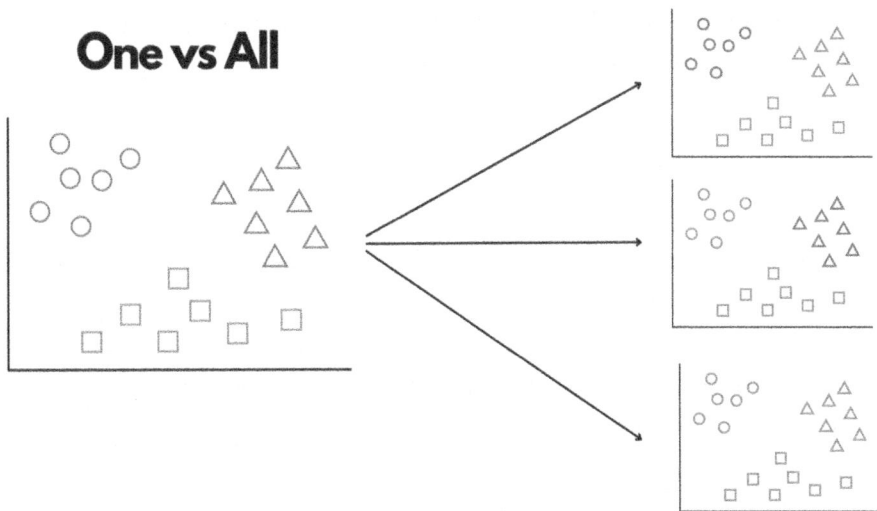

Figure 1-5. One vs. all multiclass strategy

Caution It is important to note that if you have many classes with many samples each, there is a risk that each classifier is trained on an imbalanced dataset (few samples that belong to one class and many samples that belong to the other class), which can lead to poor results.

One vs. One

The *one vs. one* strategy involves training a separate binary classifier for each pair of classes. In a three-class problem (A, B, C), the one vs. one strategy would involve training three binary classifiers.

- Classifier 1: A vs. B
- Classifier 2: A vs. C
- Classifier 3: B vs. C

19

CHAPTER 1   TINY MACHINE LEARNING

The number of classifiers grows quadratically with the number of classes according to the following formula.

$$numberofclassifiers = \frac{numberofclasses \times (numberofclasses - 1)}{2}$$

For example, you will need 45 classifiers for a classification task with 10 labels. The one vs. one strategy does not require that a classifier produces a score for its prediction. The winning class has the most votes if the output is a discrete 0 or 1 (see Figure 1-6). If the classifier can produce a confidence score, you may sum either the scores or the votes.

As with the one vs. all strategy, it may happen that no class achieves the majority of votes, and the sample should be marked as *unknown* or *unclassifiable*.

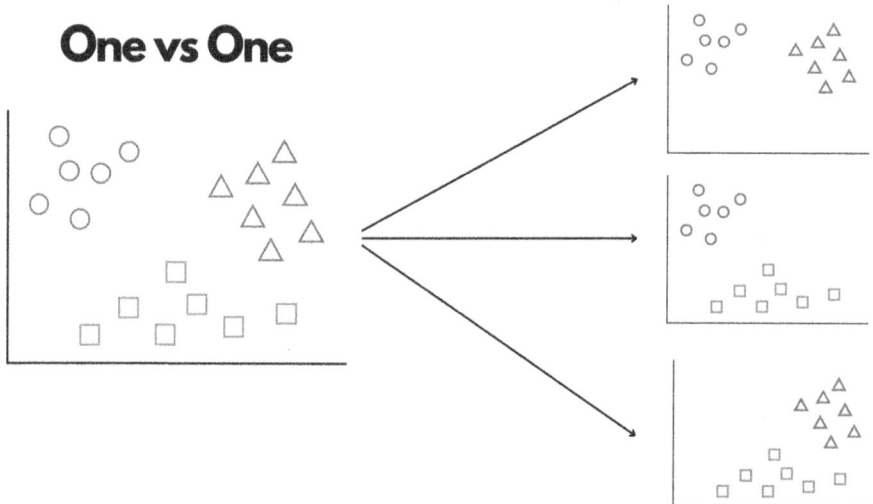

Figure 1-6. *One vs. one multiclass strategy*

Metrics

To evaluate the performance of a model, you need a metric. In supervised machine learning, a metric is a numerical value that measures how well the model's output matches the expected output. For classification problems, a metric tells you the number of classifications the model got correct from the total number of predictions. For regression problems, it measures the difference between the predicted and expected values.

Metrics for Binary Classification

In binary classification, the outcomes are limited to 0 and 1. A simple and widely used metric is *accuracy*, which is defined as the number of correct predictions divided by the total number of predictions. A higher accuracy indicates better performance. However, accuracy can be misleading, as it does not provide information about the model's performance in each class.

To illustrate the fallacy of accuracy as a metric, let's consider an example. Suppose you train a spam detector to classify emails as spam or not spam (a.k.a. "ham"). You have 900 ham emails and 100 spam emails in your dataset. The model achieves an accuracy of 90%. Can this be considered a satisfactory result? Actually, no, since 90% is the accuracy you get by always classifying an email as ham (which means the classifier didn't learnt anything at all!).

Two additional metrics are commonly used to address this limitation: *precision* and *recall*. To understand what they represent, let's first enumerate the possible outcomes in a binary classification problem.

- True positive (TP): The true outcome was 1, and the classifier correctly predicted 1.

- True negative (TN): The true outcome was 0, and the classifier correctly predicted 0.

CHAPTER 1 TINY MACHINE LEARNING

- False positive (FP): The true outcome was 0, and the classifier wrongly predicted 1.

- False negative (FN): The true outcome was 1, and the classifier wrongly predicted 0.

Precision measures how many times the classifier was correct when it predicted 1. *Recall* measures how many times the classifier predicted 1 against the total number of actual 1s.

$$precision = \frac{TP}{TP + FP}$$

$$recall = \frac{TP}{TP + FN}$$

Figure 1-7 explains the same concepts with a diagram.

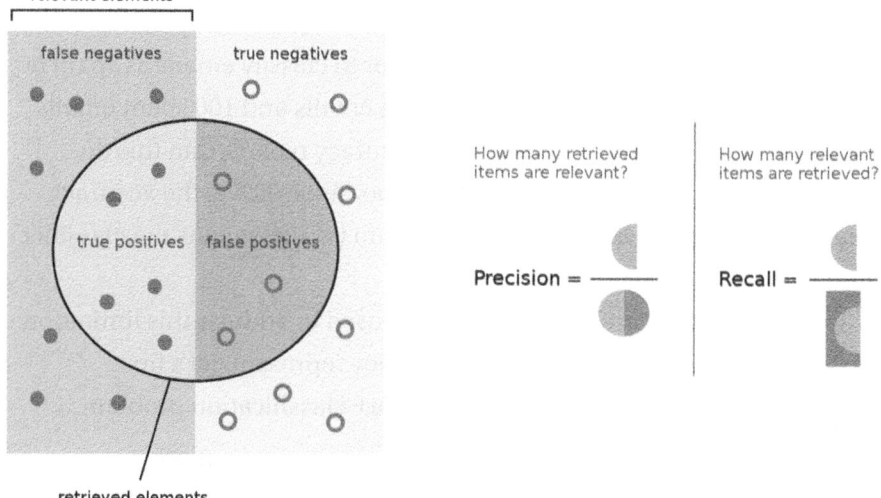

Figure 1-7. *Precision and recall*

CHAPTER 1   TINY MACHINE LEARNING

Let's go back to the spam detector example that always classifies emails as ham. Assign the value 0 to "ham" and 1 to "spam". Let's compute the metrics.

```
True positive  = 0      How many correct spam predictions?
True negative  = 900    How many correct ham predictions?
False positive = 0      How many wrong spam predictions?
False negative = 100    How many wrong ham predictions?

Precision = TP / (TP + FP) = 0 / 900 = 0%
Recall    = TP / (TP + FN) = 0 / 100 = 0%
```

We now have a completely different view of our classifier performance. Our classifier obviously did an awful job in learning how to recognize a spam email, even if the overall accuracy looked high.

How can it be possible that the classification effectiveness is so low given an accuracy so high? It happened because the dataset is *unbalanced*: the number of instances of one class (ham) is much larger than that of the other class (spam). In these cases, many algorithms favor the majority class due to the implementation details of how they learn (they optimize for accuracy instead of precision and recall).

If you prefer a single metric to describe your classifier that doesn't fool you as accuracy does, you can refer to the *F1 score*, which combines precision and recall into a single number.

$$F1 = \frac{2 * precision \times recall}{precision + recall}$$

A high F1 score means that both precision and recall are high; a low F1 score needs more investigation since it can be due to either low precision or low recall (or both!).

CHAPTER 1 TINY MACHINE LEARNING

Metrics for Multiclass Classification

In multiclass classification, you can use the same binary metrics (accuracy, precision, recall) for each class, applying a one vs. all binarization scheme. However, a more informative metric is the confusion matrix (see Figure 1-8).

A confusion matrix is a table that displays the number of correct and incorrect predictions for each class. The table has rows and columns equal to the number of classes, where each cell represents the number of samples whose true class is the row label and whose predicted class is the column label.

Figure 1-8 depicts an example of a confusion matrix referred to the Iris flower dataset [1], which tries to assign the correct Iris species (*setosa*, *virginica*, and *versicolor*) based on sepal and petal dimensions.

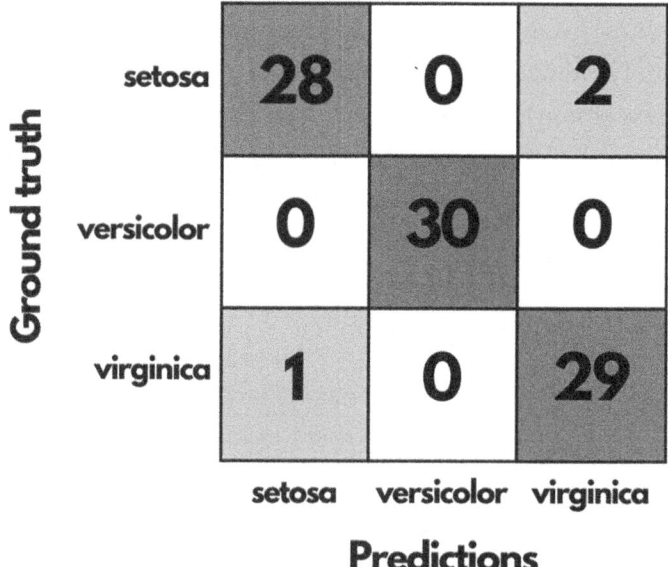

Figure 1-8. Example of confusion matrix for the Iris flower dataset

CHAPTER 1 TINY MACHINE LEARNING

The diagonal cells of the matrix represent the number of correct predictions, while the off-diagonal cells represent the errors. The confusion matrix provides valuable insights into the model's performance, such as the following.

- The overall accuracy of the model (sum of left diagonal against the total number of samples)
- The precision and recall of each class
- The number of misclassified samples

In Figure 1-8, you interpret the off-diagonal values as follows.

- The number 1 represents the samples whose true class was *versicolor* but were classified as *setosa*.
- The number 2 represents the samples whose true class was *setosa* but were classified as *virginica*.

This matrix can tell you a lot of information at a glance.

- The classifier was pretty good (72 correct classifications out of 75). Errors are low compared to correct classifications.
- *Versicolor* has 100% precision (when the classifier predicted *versicolor*, it always was a true *versicolor*) and recall (all *versicolor* samples were correctly picked).
- *Setosa* has 96% precision (28/29, the classifier was wrong once).
- *Virginica* has 96% recall (29/30 occurrences were correctly picked).

CHAPTER 1 TINY MACHINE LEARNING

This is a pretty good confusion matrix. (The Iris flower dataset is a toy dataset, and it is very easy to classify.) You may encounter much worse cases in your real-world projects, as in Figure 1-9.

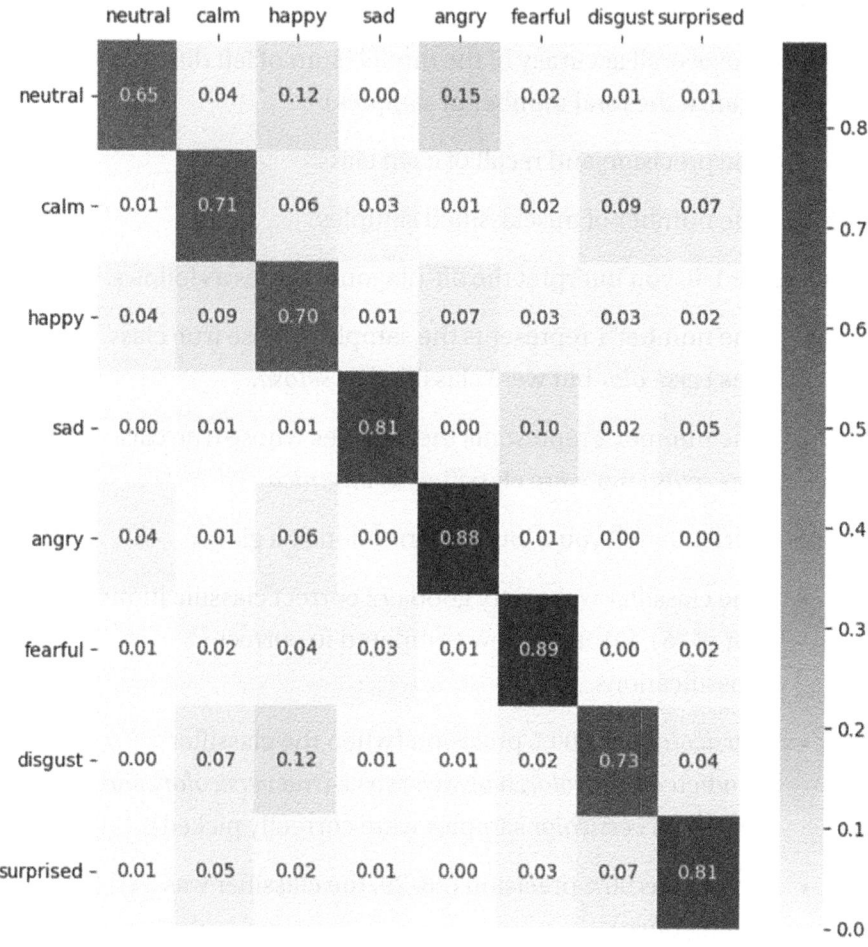

Figure 1-9. *Poor confusion matrix 1*

In this example, the overall accuracy is in the 70%–80% range, and you can see that the errors are uniformly spread across all classes. In these cases, you need to increase the modeling power of your current model via parameter tuning (if possible) or look for a more capable model.

Metrics for Regression

Regression is predicting a continuous variable from a set of inputs. In this case, you typically do not aim for a perfect match between expected and predicted values. Instead, you want your predictions to fall within an acceptable error range from the actual ones. There are two main errors that are commonly used as regression metrics.

Root Mean Squared Error

The *root mean squared error* (RMSE) is the square root of the average of the squares of the errors.

$$RMSE = \sqrt{\frac{1}{N}\sum_{i=1}^{N}(truth_i - pred_i)^2}$$

$truth_i$ is the expected value and $pred_i$ the predicted one.

Using the square operator in the RMSE gives more importance to larger errors and outliers and always produces a positive result. Taking the square root of the result has two effects: it makes it easier to compare distinct RMSE values by providing a narrower range, and it gives the RMSE the same unit as the output variable, such as centimeters for a person's height regression task.

Note A good regression model has a low RMSE.

Mean Absolute Error

The *mean absolute error* (MAE) is the average of the absolute value of the errors. It differs from the MSE because the square operator does not inflate error magnitudes, so their contribution is linear instead of quadratic. We still apply some *positivization* (absolute value) to avoid errors of opposite sign producing a null or very low error metric.

CHAPTER 1　TINY MACHINE LEARNING

$$MAE = \frac{1}{N}\sum_{i=1}^{N}|truth_i - pred_i|$$

Note　A good regression model has a low MAE.

Which one should you choose? It depends on your project, and until you are familiar with machine learning in general, I recommend you monitor both while developing.

Coefficient of Determination

The *coefficient of determination* (R^2) is a number between 0 (worst case) and 1 (best case) that expresses how well our input features explain the output. If it is high (larger than 0.9), our features contain almost all the information required to make a good prediction. If it is low, we're probably missing important features that are relevant to the output. In a real-world scenario, you can expect a good R^2 to be in the range of 0.8–0.95. Lower values may indicate that you need to find better attributes as predictors. Higher values may signal a trivial problem or a leak of the output into the inputs.

$$R^2 = 1 - \frac{\sum_{i=1}^{N}(truth_i - pred_i)^2}{\sum_{i=1}^{N}(truth_i - mean(truth))^2}$$

In the best case, $\hat{y}_i$ (the prediction) matches y_i (the expected output) perfectly, so the fraction is 0 and R^2 is 1.

Overfitting and Underfitting

Machine learning models can vary in complexity, and even the same model can be configured to be more or less descriptive. You might assume that the more descriptive a model is, the better the results. However, this is not always the case.

Overfitting is a condition where a model learns the training data too closely. In this scenario, the model begins to memorize the data instead of learning to generalize it. Several factors can cause this.

- A small and/or noisy dataset
- A dataset with many uninformative attributes
- A model with more descriptive power than required by the data

For example, let's say you want to create a model to predict a house price. You select input attributes such as the area, number of rooms, and whether it has a swimming pool. However, you also include attributes like the color of the walls, the size of the TV, and the height of the current owner. How much would you trust a model that assigns higher prices to houses whose owner is taller than 180 cm?

To address overfitting, you can try the following solutions.

- Choose a simpler model or constrain the model (if possible).
- Reduce the number of data attributes.
- Gather more data.
- Reduce noise in the data (fix errors and remove outliers).

Underfitting is the opposite of overfitting. It occurs when a model cannot learn the underlying data pattern and exhibits poor performance. This can be due to a model being too simple or the data missing important features.

For instance, you might be using a linear model to learn a quadratic relation or try to predict a house price based solely on its year of construction. To address underfitting, you can try the following solutions.

- Choose a more complex model or release its constraints (if possible).

- Increase the number of data attributes.

- Gather more data.

- Reduce noise in the data (fix errors and remove outliers).

Figure 1-10 visually shows the problem of overfitting and underfitting for classification and regression tasks.

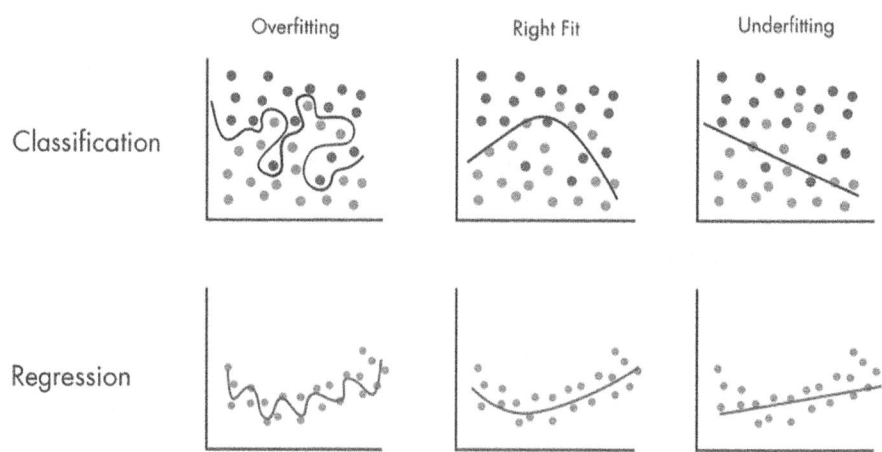

Figure 1-10. *Underfitting and overfitting in regression and classification*

Training vs. Validation vs. Test

When developing a machine learning project, it is crucial to evaluate the effectiveness of different features, models, and model parameters' configurations using a fair and objective evaluation process.

One of the most superficial errors we may commit is to stop iterating when our metrics look good, only to later discover that our model is overfitted. To avoid overfitting, where a model memorizes the input rather than learning from it, we typically split our data into three buckets: training, validation, and test sets.

The *training data* is fed as input to the algorithm to create a model. Since this is the most "data-intensive" process, the training set typically comprises 50%–70% of the samples. The model learns from this data and adjusts its parameters to minimize errors.

The model never uses the *validation data* to learn. Instead, it selects the model parameters that yield the highest metric (e.g., accuracy) on this unseen data. Since the model has not seen this data during training, it must generalize from the training samples to achieve a satisfactory result. Typically, 10%–20% of the data is used as validation. If no model parameters need to be tuned, the validation set is unnecessary.

To assess the actual model performance (after training), we use the *test set*, which is another split of the original data that the classifier has never seen during training or validation. The metric reported on this data is the value we expect to see in the real world. Typically, 10%–30% of the data is used as a test.

You may wonder why both validation and test sets are necessary. If we only use the test set for evaluation, we will select the model with the highest metric (accuracy, F1 score, MAE, RMSE, etc.) on that specific set. This result will be biased, and based on new, never-seen data, that model could (and probably will) perform worse. We are overconfident in our model's prediction power.

Let's consider an example to illustrate these concepts better. Imagine you're a teacher, and you want to test your students' knowledge. You give them a quiz to see how well they've learned the lesson's material.

The training set is like the textbook and class notes. You use this data to teach the students (your machine learning model) the concepts. They learn from this data and try to generalize the formulas or key concepts.

The validation set is like a practice quiz. You use this data to see how well the students have learned, but you can still correct their mistakes and adjust your teaching methods. This helps them fine-tune their understanding of the lessons (model's parameters) for better results.

The test set is like the final exam. You use this data to evaluate a student's knowledge in a real-world scenario without any feedback or corrections. This gives you an unbiased estimate of how well the students (model) will perform in the real world.

Summarizing, here's why you need both validation and test sets.

- **To prevent overfitting**: If you only use the training set, the model might become too specialized in the training data and not generalize well to new, unseen data. The validation set helps prevent overfitting by showing how well the model performs on unseen data.

- **To optimize parameters**: The validation set helps you tune your model's parameters. You can try different settings, evaluate the model on the validation set, and choose the best one.

- **To drop bias**: The test set provides an unbiased evaluation of your model's performance. Since you didn't use this data to train or optimize the model, it gives you a realistic estimate of how well the model will perform in real-world scenarios.

Feature Engineering

Feature engineering transforms raw data into a more effective set of inputs for machine learning models. Raw data can be noisy, redundant, or useless, and these values can have a negative impact on the model's learning process. By transforming the data into a format more suitable for the model, you can significantly improve the accuracy of predictions. Some common feature engineering steps include the following.

- Converting an audio signal into its frequency components

- Extracting statistical moments (such as mean, variance, and standard deviation) or spectral values (such as energy and entropy) from time series data

- Scaling input data into a fixed range (such as [0, 1] or [-1, 1]) can help to prevent features from dominating the model's predictions.

Each data type typically has a standard feature engineering pipeline that you can reuse across different projects. Some of these pipelines are examined in more detail in upcoming chapters.

One key point to pay attention to is the dimensionality of the data. You may think that having more data will always make it easier for a classifier to learn how to characterize each class. However, this is not always the case. In some scenarios, having too much data can actually hurt the learning process and lead to poor results. (There is a specific name for this: *curse of dimensionality*.)

Additionally, you must consider that more input data usually results in larger and slower models. In the context of TinyML, you want to have the leanest models possible to conserve precious resources when deploying the classifiers to your board. In projects dealing with high-dimensional data, it may make sense to apply *dimensionality reduction* algorithms, which combine different features into a lower set, or *feature selection*, which is the process of discarding features that do not contribute significantly to the classification result. A feature may be (almost) useless for classification/regression tasks for two reasons.

- It is highly correlated with another one, making it redundant and unnecessary.

- It does not characterize each class/output from the others, making it irrelevant.

After feature selection, your classification accuracy may drop slightly, depending on the specific dataset. Your job is to find the optimal trade-off between accuracy and resource constraints for your specific deployment. As a general rule, the more features you can discard, the better.

Types of Data

Machine learning can operate on a variety of types of data. Some types of data may be fed *as is* to machine learning models (or at least to a group of them), and some others require important feature engineering steps to help the model make sense of them. This section covers the types of data you will encounter later in the book. It doesn't cover every industry, but likely the most common ones. They are listed by increasing complexity in terms of analysis and/or information density.

Tabular Data

Tabular data refers to a type of data that is structured and organized into rows and columns, similar to a spreadsheet or a table. Imagine a table with rows and columns, where each row represents a single observation or sample, and each column represents a feature or variable. For example, a spreadsheet representing the exam outcomes of a class of students is a good example of tabular data. Each row represents a student, and each column represents their outcome in a given exam.

Here are some key characteristics of tabular data.

- **Structured data**: Tabular data is highly structured, meaning that each row and column has a specific meaning and format.
- **Rows represent samples**: Each row in the table represents a single sample or observation, such as a student, a customer, a transaction, or a product.
- **Columns represent features**: Each column in the table represents a feature or variable, such as the exam's outcome, age, income, product category, or purchase amount.
- **Homogeneous data**: The data in each column is of the same data type, such as numerical, categorical, or text.
- **Fixed schema**: The structure of the table, including the number of columns and their data types, is fixed and well-defined.
- **Row isolation**: Each row (sample) is independent of the others. You can freely rearrange the order of rows without compromising the information that each row yields.

In IoT and physical computing, data often comes from sensors. If you consider sensors' readings on their own, one at a time, that may be considered tabular data. For example, let's say you want to use a multi-gas odor sensor to detect different types of alcoholic drinks [2]. This sensor outputs the concentrations of many gases (carbon monoxide, nitrogen dioxide, ethanol, etc.) maybe every second. You don't want to monitor these values over time: you only consider the concentrations when you read the sensor. This is tabular data. More examples of tabular data that you may work with in your embedded projects include the following.

- Atmospheric measurements (temperature, humidity, pressure)
- Color and light intensity
- Distance, speed, tilt angle

At this step, feature engineering operates on single samples. It only considers transformations that are either individual (relative to the sample at hand) or global (relative to the entire population of samples).

Chapter 2 focuses on tabular data classification, while Chapter 3 discusses regression.

Time Series Data

Time series data refers to cases where time is an added dimension to the collected data. In this context, you have two sources of information.

- The measurement values at a given point in time
- How values change over time

You can still store time series data in a tabular format, but the order of the rows is crucial. If you rearrange the rows in random order, you lose a significant amount of the intrinsic patterns of the data. Time series data may not have a fixed or uniform sampling frequency, and different sensors may have different frequencies (from a few samples per second to hundreds).

The following are some key differences between time series and tabular data.

- **Time**: It is as important as measurement values. A single sample holds very little information; what makes a pattern is how the values evolve over time.
- **Numeric only**: While tabular values may be a combination of numeric and categorical data, time series data is always numeric.

CHAPTER 1   TINY MACHINE LEARNING

Given that data comes in as a stream of values and our hardware memory is limited, it is mandatory to *window* the data. Windowing is the process of only considering a chunk of data (the most recent one) at a time, discarding old data as fresh one comes in. Figure 1-11 depicts the windowing process over time.

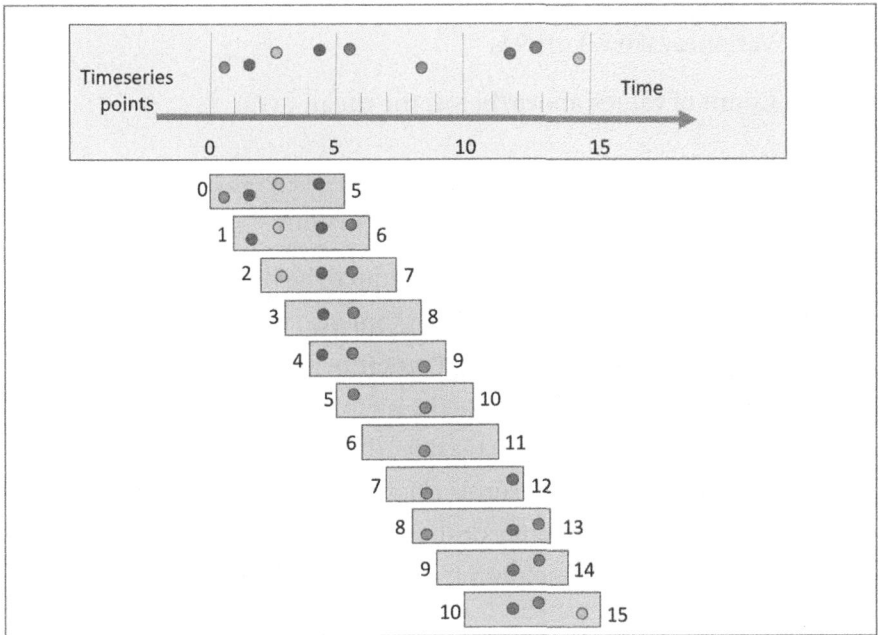

Figure 1-11. *Sliding window*

The following are some typical sources of time series data.

- accelerometer and gyroscope to detect movements or vibration patterns

- current and voltage to classify a device operating mode (e.g., idle, light load, or heavy load)

- atmospheric sensors (temperature, humidity, pressure) to predict if it will snow in the next hour

- brain and muscle electric signals (EEG and ECG) to remotely control devices

Feature engineering of time series often consists of extracting statistical values from the windows in the time domain, such as the following.

- Minimum/maximum/average
- Variance/skew/kurtosis
- Count of values above/below the mean
- Zero crossings
- The frequency domain (Fourier transform)

These steps are required not only to reduce the dimensionality of the data (which can be a problem on its own) but also to *de-noise* the input and improve its information content. Remember that each sample in a time series holds little information on its own: packing many samples together doesn't sum information up linearly.

Chapter 4 approaches time series classification using the Edge Impulse low-code platform and frequency domain features. Chapter 5 discusses time series classification using pure Python and time domain descriptors.

Audio

Audio is a special kind of time series data. In the embedded world, it works at pretty standard sampling frequencies (8, 16, or 20 kHz) and amplitudes (represented as a signed 16-bit integer, from −32768 to +32767). At these high frequencies, extracting features in the time domain becomes problematic, and even if you could, they may not be the most suitable solution.

We approach audio classification using a custom version of frequency domain features called a *Mel spectrogram*. This transform maps the audio data from the time domain to the time-frequency domain. In this new domain, you can compute specific features that will become the input for the classification model. Given the amount of computation required and the (near) real-time execution constraints, optimized algorithms have been developed to perform the Fourier transform in a timely manner, and many hardware vendors have even introduced hardware acceleration for it.

Chapter 6 implements spoken word classification using the Edge Impulse low-code platform.

Images and Spatial Data

Images are a special kind of spatial data with an intrinsic arrangement in the XYZ space. They are a grid of pixels arranged in rows and columns, and each pixel's value is as important as its position. Each pixel is strictly correlated with its neighbors: rearranging the pixels randomly along the rows or the columns would completely destroy the original image (the same way as rearranging time series did). However, spatial data is not limited to camera images. There exist a couple more examples that are common in the embedded environment.

- Thermal camera: this sensor perceives the infrared light that strictly correlates with objects' temperatures. Depending on the sensor, the resolution can be 8×8, 32×24, or 160×120. In this case, every pixel's value represents the estimated temperature.

- Time of flight arrays: time of flight is a technique used to detect distances by measuring how much time the light takes to bounce against an object and return to its source. With specialized hardware (called SPAD arrays), you can measure distances over many points in the 3D space in a single run. Resolution is still limited, with the most common sensors achieving a 4×4 or 8×8 output.

No matter what pixels represent (light intensity, temperature, or distance), spatial data may leverage a class of algorithms developed

CHAPTER 1 TINY MACHINE LEARNING

specifically for images. As long your data *looks like* an image (values are bounded to a specific domain, e.g., 0–1 or 0–255), they should work fine no matter what pixels represent.

It should be evident that a "visual" camera has resolutions much higher than the alternatives described (up to 5 MP for some microcontrollers). So, images carry much more information than an 8×8 SPAD array output. Processing that much data can take a lot of time and resources. Considering our definition of embedded hardware (2 MB of flash memory and 2 MB of RAM), image analysis must be limited to low-resolution images (96×96 being a good compromise between accuracy and resources' usage). Also, only a subset of the available models can fit our hardware constraints. Advanced deep learning models commonly used on desktop hardware (e.g., YOLO architectures [3]) are not suitable in their default, full width form.

Chapter 7 explains the object detection process: how to recognize objects of interest inside an image.

To close the book, Chapter 8 approaches the field of artificial neural networks for embedded devices with a hands-on, code-oriented introduction to the TensorFlow framework [4]. You learn four common network topologies and how/when to use them in Python (for training) and Arduino (for inference).

Summary

This chapter introduced the core idea of machine learning in general and the differentiating points that make TinyML a unique, separated niche. You got an overview of TinyML application industries, use cases, and what it still can't achieve. Finally, you were introduced to the technical terms and concepts of machine learning so that you don't get lost throughout the rest of the book and the kind of data you will be working with.

The next chapter starts with classifying tabular data, the simplest data type due to the row independence condition.

CHAPTER 2

Tabular Data Classification

Tabular data refers to data that is structured and organized into rows and columns, similar to a spreadsheet. Tabular data classification is the task of receiving input from the columns of a single row and deciding which class that row belongs to.

The order of the rows is not important since each one is self-contained and isolated from the others. Feature engineering on tabular data operates considering either a single row at a time (row-level feature engineering) or all the rows at once (dataset-level feature engineering).

This chapter replicates the fruit classification project from the Arduino blog [1] that detects fruit based on its color components (see Figure 2-1). However, we won't use TensorFlow or neural networks. This chapter, Chapter 3, and Chapter 5 are designed to use *traditional* machine learning, instead of deep learning (TensorFlow and similar). The models generated work on almost any microcontroller, even 8-bit ones (e.g., ATMega or Attiny series), with as little as 10–20 KB of RAM (depending on the model size).

***Figure 2-1.** Demo of the result of Chapter 2 project*

Let's go through the following steps to complete our project.

1. Capture data using our microcontroller and external sensors.

2. Load and inspect the data using Python.

3. Perform feature engineering using Python.

4. Train a classification model using Python.

5. Convert the model to C++ and deploy it back to our microcontroller.

Each step gradually introduces the theory and tools that are required (or that make it easier) to accomplish the task. Do not skip any step; otherwise, you won't be able to get the correct results at the end. The same workflow also applies to Chapter 3 and Chapter 5.

CHAPTER 2 TABULAR DATA CLASSIFICATION

Required Hardware

We need a color sensor since we want to classify different fruits based on their color. One of the following setups will work fine.

- Arduino Nano BLE Sense, with built-in APDS9960 sensor (see Figure 2-2)

Figure 2-2. *Arduino Nano BLE Sense*

- Any microcontroller with an external TCS3200 sensor (see Figure 2-3). This sensor has four control pins (S0/S1 for frequency scaling, S2/S3 for color) that need to be connected to digital output pins and one output pin for the detected signal (to be read using the `pulseIn` function).

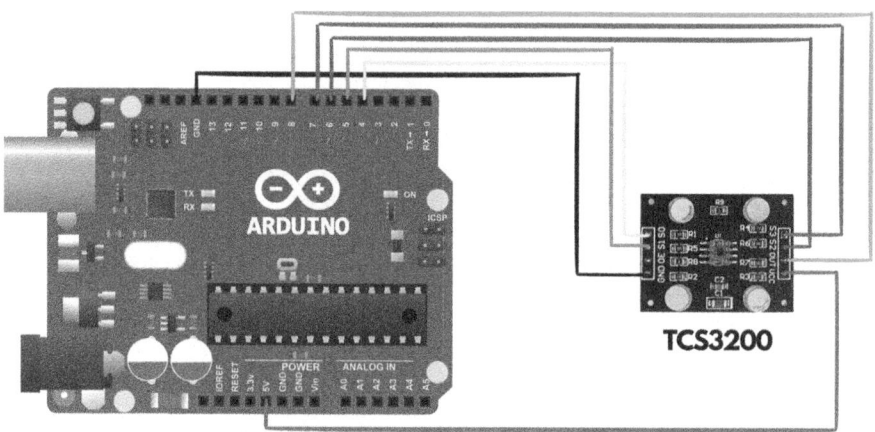

Figure 2-3. *Wiring of TCS3200 color sensor*

If you don't have fruits of different colors (e.g., banana, apple, orange), you can replace them with any colored object. *Make sure that each object has a clear, distinctive color (avoid using a banana and a lemon).*

Required Software

The only strict requirements for this project (and for all the projects in this book) are that you must install Python 3.10 or higher on your PC and the Arduino IDE (Platform IO with Arduino support is fine, too).

There are some optional steps that I suggest you carry on to ensure a smooth programming experience throughout the rest of this book, though.

Create A Python Virtual Environment

Let's use Python with some third-party packages to perform our machine learning tasks. To keep our code and dependencies isolated, we will create a Python virtual environment.

A virtual environment in the context of Python is an isolated, project-specific installation of a Python interpreter, plus all the required additional libraries that the project needs. Since you likely have many Python projects on your PC, created at different times, version compatibility problems may arise if you install packages globally (due to updates in the dependencies). A virtual environment for each project prevents these conflicts by using different versions for different projects.

A virtual environment is created only once. Then, to be used, it needs to be *activated*. Every time you close your terminal, the environment is automatically *deactivated*, and you must reactivate it the next time you want to use it.

To create a virtual environment for this book, open your terminal into a dedicated folder and run the following command.

```
# create a new virtual environment (run only once)
$ python -m venv <name of the environment>
```

To activate the environment (while inside the root folder), run one of the following (depending on your setup).

```
# Windows Command Prompt
$ <name of the environment>\Scripts\activate.bat

# Windows PowerShell
$ <name of the environment>\Scripts\activate.ps1

# Linux/OS X
$ source <name of the environment>/bin/activate
```

These commands translate to the following if you want the virtual environment to be named *tinyml*.

```
$ python -m venv tinyml

# Windows Command Prompt
$ tinyml\Scripts\activate.bat

# Windows PowerShell
$ tinyml\Scripts\activate.ps1

# Linux/OS X
$ source tinyml/bin/activate
```

After activating the virtual environment, you can start installing the required packages. I created a companion package for this book called tinyml4all. To install, execute the following line inside your virtual environment.

```
(tinyml)$ python -m pip install tinyml4all
```

Tip If the Python command is not recognized, try to replace it with python3.

When running the Python code examples from this book, be sure you have activated your venv first; otherwise, you get many errors (missing dependencies, most of the time).

---

tinyml4all Arduino Library

To assist you in the tasks that run on the microcontroller, I created a companion Arduino library called tinyml4all that you can install from the Arduino Library Manager. Even though it is completely optional, I strongly

CHAPTER 2 TABULAR DATA CLASSIFICATION

suggest you install it. This way, you can run the sketches provided in this book without any modification. Figure 2-4 shows the correct library from the Arduino Library Manager window.

> **Tip** To open the Library Manager, open the Arduino IDE and navigate to Sketch ➤ Include library ➤ Manager libraries.

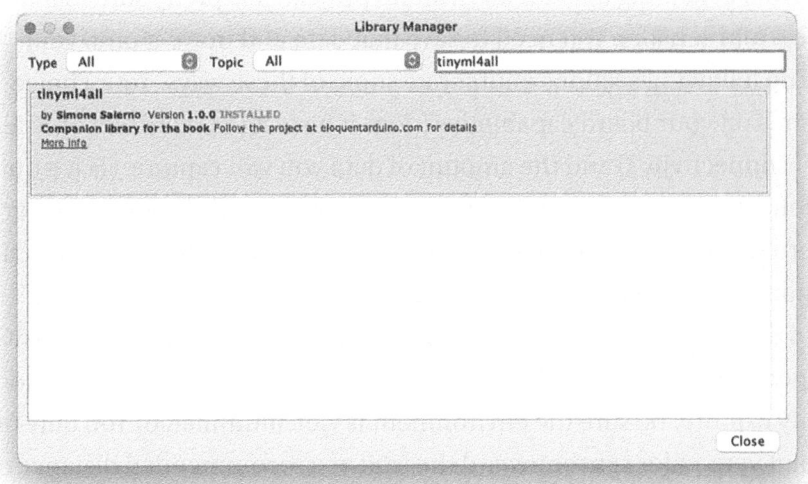

Figure 2-4. *Install tinyml4all Arduino library from the Library Manager*

Capture Data

The first step in any machine learning project is to collect data. Sometimes, you may already have data (collected in the past or downloaded from the Internet). Other times, you need to collect new data from your board and sensors.

47

Collecting quality data is a crucial step. There's a saying in the machine learning industry that says, "garbage in, garbage out." If your input data is noisy, missing, and low quality, your model's output mirrors that quality. So, take any effort you can to collect the best data possible. Often, collecting data is time-consuming or costly. Depending on your project requirements, you must evaluate how much value you assign to high-quality input data. Not paying attention to this step can invalidate all later processing and preclude a successful result.

When collecting data for an Arduino project, it is pretty sure that data comes from sensors. You need to save that data and move it onto your PC. The following sections enumerate some of these ways. Your choice depends on your board capabilities (does it have an SD card slot or BLE/Wi-Fi connectivity?) and the amount of data you will capture. (Is it a quick session, or do you collect for many minutes or hours?) If you have other preferred methods, you can use them. What matters is that you have a file (or many files) on your PC containing the collected data.

In the context of this project, our data is the red, green, and blue (RGB) components of the light reflected by each of the fruits. To achieve a good quality capture, be sure the environment is well illuminated. You only need to point your color sensor toward the fruit at a recommended distance of 15–30 cm (see Figure 2-5).

CHAPTER 2   TABULAR DATA CLASSIFICATION

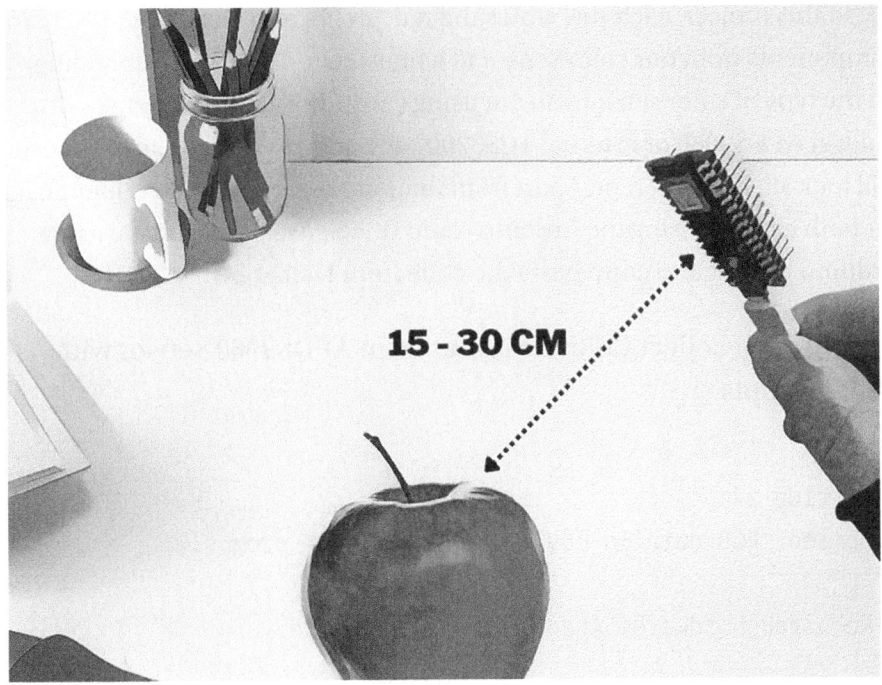

Figure 2-5. Example setup for RGB collection

1. Manually Copy Data from Serial Monitor

It may sound dumb, but the most straightforward, simple way to get data from your microcontroller is to print it on the Serial Monitor and manually copy and paste it into a file on your PC. For quick-and-dirty projects, this is by far the fastest way to get started.

Of course, you want to properly format your data before printing otherwise, you won't be able to easily process it in Python later. If you installed the `tinyml4all` Arduino library, this should be fairly easy. The most suitable format for Python ingestion and processing is comma-separated value (CSV) encoding. It is a text-based format that stores data in rows and columns. Each row is on a line. Each column is separated from the others with a comma.

CHAPTER 2   TABULAR DATA CLASSIFICATION

In this project, each row stores the red, green, and blue light components from our color sensor at a one-second interval. Depending on the type of color sensor you are using (Arduino Nano BLE Sense with built-in APDS9960 or external TCS3200), the code to instantiate the sensor will look slightly different. Apart from that, the rest of the code is identical for both cases. If using the Arduino Nano BLE Sense board, create a new Arduino project and copy-paste the code from Listing 2-1.

Listing 2-1. Collect Color Readings from APDS9960 Sensor with User Prompts

```
/**
 * Listing 2-1
 * Collect RGB data in CSV format from user prompt.
 *
 * Required hardware: Arduino Nano BLE Sense.
 */
#include <Arduino_APDS9960.h>
#include <tinyml4all.h>

using tinyml4all::promptString;
using tinyml4all::promptInt;
using tinyml4all::printCSV;

tinyml4all::APDS9960 sensor;

void setup() {
    Serial.begin(115200);
    while (!Serial);
    Serial.println("Collect RGB values as CSV");

    // init sensor (will throw an error if it fails)
    sensor.begin();
}
```

CHAPTER 2 TABULAR DATA CLASSIFICATION

```
void loop() {
    // get fruit name and number of samples from user
    String fruit    = promptString("Which fruit is this?");
    int numSamples = promptInt("How many samples to capture?");

    for (int i = 0; i < numSamples; i++) {
        // read sensor values and print in CSV format
        sensor.readColor();
        printCSV(sensor.r, sensor.g, sensor.b, fruit);
        delay(1000);
    }
}
```

Note If you're using an external TCS3200 sensor, refer to the book's code repository.

Flash the sketch to your board and open the Serial Monitor. Choose a fruit, point the board toward it at 15–30 cm, and enter its name when prompted; then enter 50 for the number of samples. The CSV lines appear every second (see Figure 2-6).

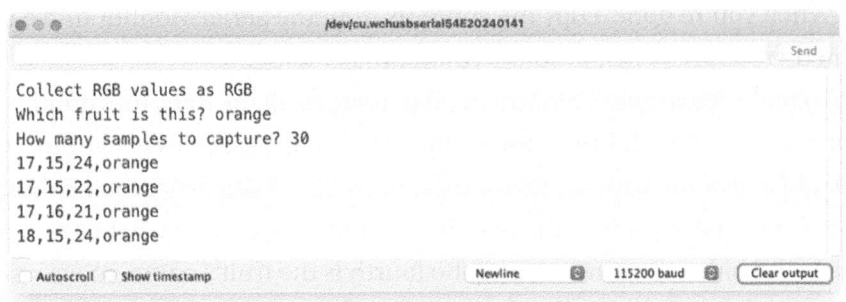

Figure 2-6. *Serial output of fruit data collection*

To collect good quality and robust data, move the sensor around a bit while the collection procedure is in progress—get closer, then further, then to the left, then to the right (while still pointing in the direction of the fruit)—to create some variability in the data (see Figure 2-7). Repeat the same process for each fruit you're going to recognize. I suggest you collect 30-50 samples for each.

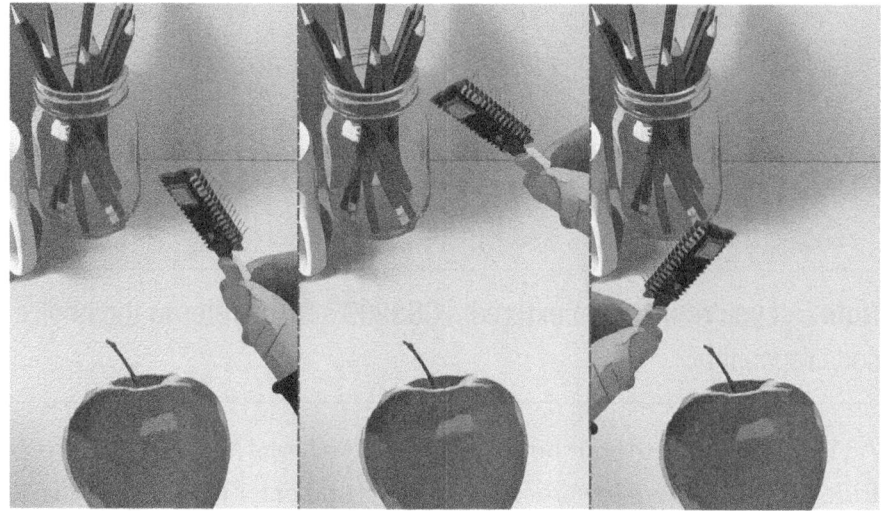

Figure 2-7. *Collect RGB values from different angles*

When you're done, copy the contents from the Serial Monitor into a file named fruits.csv inside your Python project folder.

To make it a *proper* CSV, you need to remove all the lines that don't contain data ("Which fruit is this?" and "How many samples to capture") and add a heading with the columns' names, specifying what each column represents. In this project, the first three values represent the light's red, green, and blue components, and the fourth is the fruit's name, so we prepend a line with the contents r, g, b, fruit.

```
# excerpt of the file fruits.csv
r,g,b,fruit
17,12,9,banana
```

CHAPTER 2 TABULAR DATA CLASSIFICATION

17,13,8,banana
16,13,9,banana
18,10,10,banana
38,23,18,orange
40,22,18,orange
49,25,19,orange
65,31,22,orange
54,25,18,orange

2. Read Serial Output from Python

Manually copy-pasting data from the Serial Monitor works fine until you have a low volume of data. If your data is produced at high rates or has many attributes, or you need to collect it for a long time, it may be more practical to use a Python script that saves you from the manual work and automatically reads the Serial output from the microcontroller, saving it to a file.

The Arduino code in Listing 2-2 has been updated to continuously read and print values to the Serial output without human intervention. We handle the prompting from Python.

Listing 2-2. Collect Color Readings from APDS9960 Sensor Without User Prompts

```
/**
 * Listing 2-2
 * Collect RGB data in CSV format without user prompt.
 *
 * Required hardware: Arduino Nano BLE Sense.
 */
#include <Arduino_APDS9960.h>
#include <tinyml4all.h>
```

CHAPTER 2 TABULAR DATA CLASSIFICATION

```
using tinyml4all::promptString;
using tinyml4all::promptInt;
using tinyml4all::printCSV;

tinyml4all::APDS9960 sensor;

void setup() {
  Serial.begin(115200);
  while (!Serial);
  Serial.println("Collect fruits colors as CSV");

  // init sensor (will throw an error if it fails)
  sensor.begin();
}

void loop() {
  // read sensor values and print in CSV format
  // without user intervention
  sensor.readColor();
  printCSV(sensor.r, sensor.g, sensor.b);
  delay(1000);
}
```

Create a new Python script inside the project's folder named capture_colors.py, with the contents from Listing 2-3. The script asks you to input the fruit name and the number of samples to capture, and then the automatic acquisition over the serial port starts. When you've finished, leave the fruit name blank to exit.

Caution Be sure to replace the highlighted values in the script with your own values!

CHAPTER 2 TABULAR DATA CLASSIFICATION

Listing 2-3. Capture Data from Serial in Python

```python
from tinyml4all.tabular import capture_serial

while True:
    # prompt user for fruit name and number of samples
    fruit = input("Which fruit is this? ")
    num_samples = input("How many samples to capture? ")

    # exit when fruit or number of samples is blank
    if not fruit or not num_samples:
        break

    # start the capturing
    # will connect to the serial port and read its data
    capture_serial(
        # board serial port
        # * is a wildcard match
        # on Windows, this will look like COM1 or similar
        port="/dev/cu.usb*",   #
        # must match with the Arduino sketch
        baudrate=115200,
        # file name where output will be stored
        save_to="fruits.csv",
        # the list of columns to save
        headings="r, g, b, fruit",,
        # board only sends r, g, b
        # so we append the fruit manually
        append_values=[fruit],
        num_samples=int(num_samples)
    )
```

To run the code, open a terminal inside the folder where the script is located and run the following command. (**remember to activate the virtual environment first!**)

CHAPTER 2   TABULAR DATA CLASSIFICATION

```
(tinyml)$ python capture_colors.py
Which fruit is this? banana
How many samples to capture? 30
Connected to serial port
100%|███████████████████████| 30/30 [00:30<00:00, 1.01s/it]
Disconnected from serial port
```

Repeat the process for every fruit you want to classify. When done, if you open the fruits.csv file, you find content similar to the following.

```
# excerpt of the file fruits.csv
r,g,b,fruit
38,23,18,banana
40,22,18,banana
49,25,19,banana
65,31,22,banana
54,25,18,banana
```

> **Tip** If you get an error reading the Serial port, be sure that you entered the correct port name and you don't have the Arduino IDE Serial Monitor open already! If that's the case, close it and re-run the Python script.

3. Save to SD Card

If you don't want wires going from your board to your PC, or the board cannot be easily reached, you can store data on an SD card. Not all boards come equipped with an SD slot, so consider this when buying a new board. Otherwise, you can add an external SD card reader connected via SPI pins. Listing 2-4 is a sketch that stores the RGB readings of the Arduino

CHAPTER 2 TABULAR DATA CLASSIFICATION

Nano BLE Sense to an externally connected SD card reader. Many SD card readers require a connection to the default SPI pins of the board (check your board's datasheet) plus one more pin (called CS) that can be connected to a pin of your choice. That value must be configured in Listing 2-4 to make it work.

> **Caution**   Replace the values in bold with your specific values!

Listing 2-4. Save Color Readings to SD Card

```
/**
* Listing 2-4
* Collect RGB data in CSV format and store on SD card.
*
* Required hardware: Arduino Nano BLE Sense.
* Required hardware: SPI SD card reader
*/
#include <SPI.h>
#include <SD.h>
#include <Arduino_APDS9960.h>
#include <tinyml4all.h>

tinyml4all::APDS9960 sensor;
tinyml4all::SDCard card;
// replace with the correct pin!
// see the SD card reader module datasheet to find this value
const uint8_t CS_PIN = 4;

void setup() {
  Serial.begin(115200);
  while (!Serial);
  Serial.println("Collect fruits colors as CSV on SD card");
```

```
  // init sensor and SD card
  // then open file for writing
  // will throw an error if something goes wrong
  sensor.begin();
  card.begin(CS_PIN);
  card.writing("fruits.csv");
}
void loop() {
  // read sensor values and print in CSV format
  sensor.readColor();
  card.println(sensor.r, sensor.g, sensor.b);
  delay(1000);
}
```

Now, you can power the board using a power bank (Figure 2-8) or a battery and collect data without strings attached. Since we can't interact with the board, we're collecting a single file with only the RGB components, without the fruit name. Collect a few seconds of data for each fruit, then somehow mark the end of the fruit's session (e.g., put your hand in front of the sensor so that it reads very low values for all the components) and move to the next.

CHAPTER 2 TABULAR DATA CLASSIFICATION

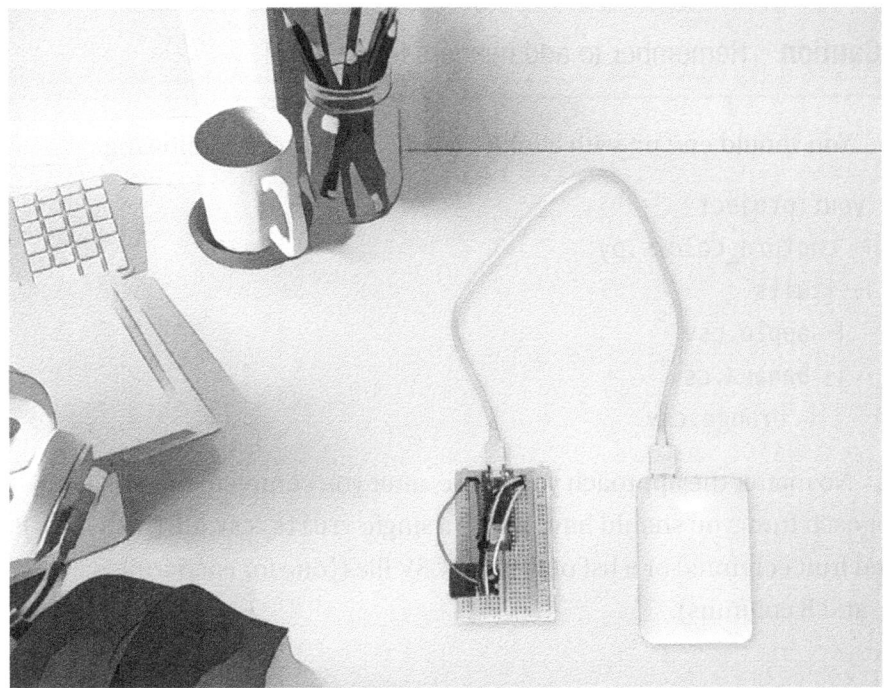

Figure 2-8. External SD card reader setup

Tip If there's a lag between the moment you power up the board and the moment you start pointing the sensor to the fruit, delete the first few lines. Do the same with the last few ones.

When you're done, move the file from the SD card to the project folder. You must impute the correct fruit to each row. One way is to open the file in Excel and manually add a fruit column (use the drag option to edit cells in bulk) and delete the spurious rows (those with all low values).

Another way is to create a file for each fruit: copy-paste all the rows relative to the same fruit into their own CSV file. (I suggest you create a dedicated folder to put them in.)

> **Caution** Remember to add headers to each file!

You should end up with a folder layout similar to the following.

```
|- your project
  |- capture_colors.py
  |- fruits
    |- apple.csv
    |- banana.csv
      |- orange.csv
```

No matter the approach you chose, after you complete the procedure for each fruit, you should have either a single `fruits.csv` file (with R, G, B, and fruit columns) or a list of distinct CSV files (one for each fruit, with R, G, and B columns).

Load and Inspect Data

The next step in our tiny machine learning workflow is to load and inspect data. You should never underestimate the value of visual inspection because you are the first sentinel to judge whether your data is garbage. Blindly running machine learning algorithms on data that you know nothing about results in poor performance at best or wasted work at worst.

The next section lists some ways to load data in your Python script. It covers cases where your data is organized in one of two possible layouts.

- One CSV file for each class, stored inside the same folder
- One file with all the data and labels

In the `tinyml4all` package, each type of data has a specialized class responsible for loading and manipulation: in the case of tabular data, that class is named `Table`.

CHAPTER 2 TABULAR DATA CLASSIFICATION

Caution Regardless of how you stored the files, it is mandatory that the first line of each CSV file is the columns' headers!

One File per Class

In this scenario, the files only contain observations (RGB light values) but not labels. The label is encoded implicitly in the file name. Use Listing 2-5 to load data arranged in this format.

Listing 2-5. Load All CSV Files from a Folder

```
from tinyml4all.tabular.classification import Table

table = Table.read_csv_folder("fruits")

# print the first few rows of table
print(table.head())
```

	r	g	b	__target_name__
0	38	23	18	orange
1	40	22	18	orange
2	49	25	19	orange
3	65	31	22	orange
4	54	25	18	orange

```
# print a few metrics for the table
print(table.describe())
```

61

	r	g	b
count	150	150	150
mean	33,3	19,9	16,9
std	12,24	5,6	3,86
min	12	8	8
25%	22	16,25	15
50%	33	19	17
75%	42,5	23	19
max	65	59	46

The preceding table is for a dataset of three files with 50 lines each (so count = 150). It is a quick summary of the distribution of your data.

Tip If you get a FileNotFound error or something similar, ensure the path to the folder is correct and contains valid CSV files.

One File for All Classes

If your data is packed into a single file, refer to Listing 2-6.

In this case, you must manually specify which column contains the ground truth labels.

Listing 2-6. Load a Single CSV File

```
from tinyml4all.tabular.classification import Table

table = Table.read_csv("fruits.csv")
# specify column that contains the labels
table.set_targets(column="fruit")
```

	r	g	b
count	150	150	150
mean	33,3	19,9	16,9
std	12,24	5,6	3,86
min	12	8	8
25%	22	16,25	15
50%	33	19	17
75%	42,5	23	19
max	65	59	46

The result is the same as earlier.

Manipulate Table

If you've ever used the pandas Python package [2], you'll notice that the table object looks like a DataFrame. And you'd be correct since the Table class mimics that API while adding new functionalities for machine learning. For those who've never used pandas, here's a list of the most common operations that you can perform on a Table instance.

Select a Single Column

Typically, your data comprises many columns (r, g, b in our project). If you want to access a single column from the table, you can use the bracket notation (see Listing 2-7).

Listing 2-7. Extract a Single Column from a Table

```
red = table["r"]
# red is a pandas.Series object
     red
```

CHAPTER 2 TABULAR DATA CLASSIFICATION

```
0        38
1        40
2        49
3        65
4        54
```

Select a Subset of Columns

In the same way you can select a single column, you can also select many columns simultaneously to create a subset table (see Listing 2-8).

Listing 2-8. Create a New Table with a Subset of Columns

```
table2 = table[["r", "g"]]
    r   g
0  38  23
1  40  22
2  49  25
3  65  31
4  54  25
```

Select a Subset of Rows

Instead of selecting a subset of columns, you may select a subset of rows based on their index (see Listing 2-9).

Listing 2-9. Create a New Table with a Subset of the Rows

```
# get a single row
row = table[0]
{'r': 38, 'g': 23, 'b': 18, '__target__': 'orange'}

# get rows from 10 to 30
```

```
table2 = table[10:30]
    r   g   b
0   40  22  18
1   49  25  19
2   65  31  22
3   54  25  18
...
19  45  23  15
```

Access the Underlying DataFrame

If you know pandas and want to access the underlying DataFrame that is wrapped inside the table object directly, you can do so by accessing its df attribute (see Listing 2-10).

Listing 2-10. Access the Underlying DataFrame Object

```
df = table.df
```

Caution The returned DataFrame is read-only! Altering it won't alter the table instance.

Apply labels

Your dataset must always have labels to perform classification. If you save each class's data inside a distinct file, the label is encoded in the file name. If you have a single file with the observations from all the classes, you should have a column that specifies the label for each row. If you don't have this column, you *must* add the labels manually.

No script does this for you automatically. You have to perform this task by hand. The only way you can speed this process up (instead of going one row at a time and inputting the correct label one by one) is if your rows are sorted by label. If you know that rows from a to b share the same label, you can refer to Listing 2-11 for editing the label in bulk.

Listing 2-11. Set Table Labels Based on Row Indices

```
# set the "orange" label on rows from 0 to 29
table.set_targets(label="orange", rows=(0, 29))

# set the "banana" label on rows from 30 to 59
table.set_targets(label="banana", rows=(30, 59))
```

Plot Data

To visually inspect your data, you can plot it. This helps you get an overview of the data at a glance. Don't forget that your job, in this early step, is to spot signs of low-quality input (values that are missing, outliers, or plain wrong).

Visually inspecting your data is an essential practice that should never be underestimated. It's crucial to remember that machine learning isn't akin to black magic; it can only distinguish patterns that genuinely exist in the dataset. If you cannot visually discern differences between classes when plotting your data, it's likely that a machine learning model struggles as well—no matter how sophisticated the model might be. Two primary types of plots are available for tabular data.

- **scatter plot of dimensionality-reduced data**: Unless your data has exactly two features (which is very unlikely), this kind of plot collapses the n columns of the data into two components to be plotted on a 2D scatter plot.

- **pair plot**: It generates a matrix of "binary" scatter plots, where each plot only considers two out of the n features.

2D Scatter Plot

Your data is composed of several columns, usually more than two. How is it possible to draw so many dimensions on a 2D plot? One obvious solution would be to arbitrarily select only two to plot and ignore the others. Of course, this discards a lot of information that is indeed present in the original data, so let's hope better solutions exist. And they do indeed.

The transformation from a high-dimensional space to a lower-dimensional one is called *dimensionality reduction*. Depending on the use case, many different algorithms exist that perform this task, each with pros and cons.

The `tinyml4all` library implements one called T-Distributed Stochastic Neighbor Embedding (t-SNE) [3] that works well for visualization purposes. You won't even need to bother since it is handled transparently for you (see Listing 2-12 and Figure 2-9).

Listing 2-12. Draw a 2D Scatter Plot of the Entire Table

```
table.scatter()
```

CHAPTER 2 TABULAR DATA CLASSIFICATION

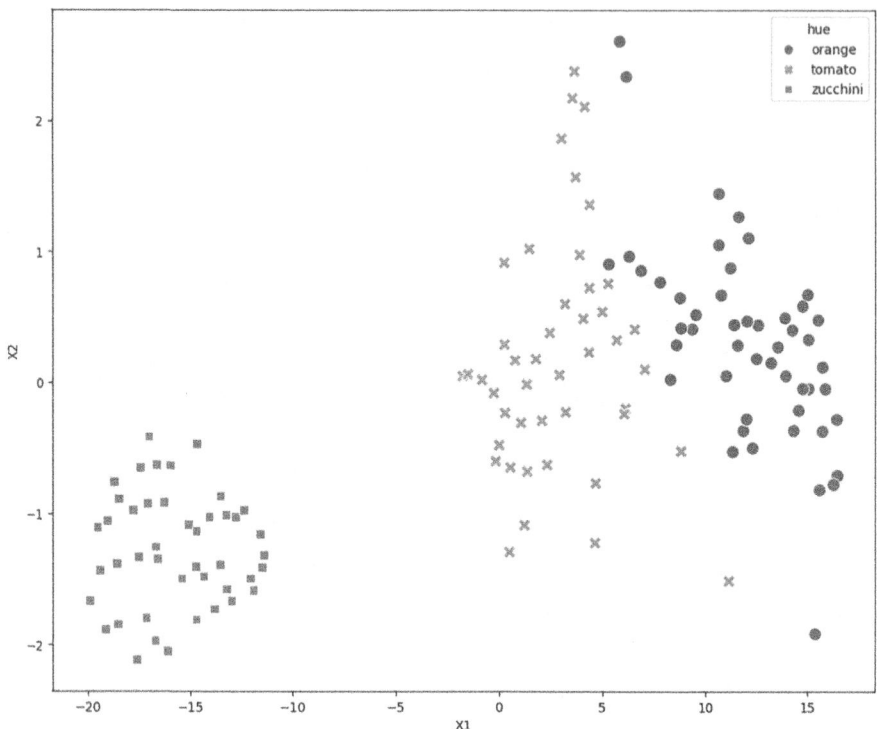

Figure 2-9. *Scatter plot of tabular dataset*

You can see that points of different fruits are pretty well clustered. This is a good sign that our data is separable and easily classified.

Pair Plot Matrix

A pair plot is a matrix of scatter plots. Instead of collapsing all the dimensions down to two, a pair plot—as the name suggests—picks pairs of features and only plots that pair (see Figure 2-10). By enumerating every single pair, the pair plot gives a detailed view of your data. It comes naturally that, with the number of scatter plots growing quadratically with the number of columns, this visualization may become slow with large datasets and hard to fit into the screen.

Nevertheless, the value you get from such visualization is worth the time. Listing 2-13 shows how to draw a pair plot for a tabular dataset.

Listing 2-13. Draw a Pair Plot of the Table

```
table.pairplot()
```

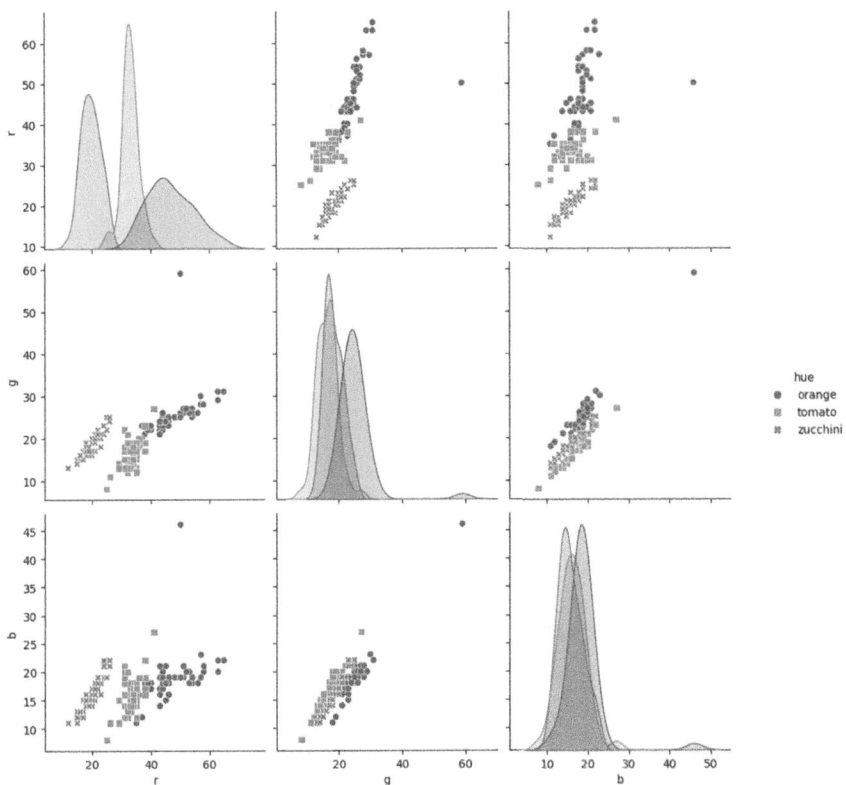

Figure 2-10. *Pair plot of tabular data*

Table 2-1 summarizes the differences between the two kinds of plots.

CHAPTER 2 TABULAR DATA CLASSIFICATION

Table 2-1. *Scatter Plot vs. Pair Plot*

	Pros	**Cons**
Scatter	Gives a quick view of the entire dataset; it is fast	A lot of information may get lost in the process, especially with high-dimensional data
Pair	Covers all the features; it is easier to spot class boundaries	Can be slow with high-dimensional or large datasets

Listing 2-14 summarizes what you've seen so far in terms of tabular data loading and visualization.

Listing 2-14. Summary of Tabular Data Operations

```
from tinyml4all.tabular.classification import Table

# load table from file
table = Table.read_csv("fruits.csv")

# load table from folder of files
table = Table.read_csv_folder("fruits")

# set labels from column
table.set_targets(column="fruit")

# set labels on a subset of rows
table.set_targets(label="apple", rows=(0, 30))

# select a single column
red = table["r"]

# select a subset of columns
rg = table[["r", "g"]]

# select a single row
```

```
row = table[0]

# select a subset of rows
first_10 = table[0:10]

# draw scatter plot
table.scatter()

# draw pair plot
table.pairplot()
```

Feature Engineering

The distinctive trait of tabular data is that every sample is totally independent of the others, so it can be treated individually and in isolation (compared to time series/audio data where each sample is correlated *temporally* with its neighbors or to images where each pixel is correlated *spatially* with the others).

Tabular data, though, can group values on different scales within the same sample. Let's consider a weather station device that collects temperature, humidity, and pressure: temperature may range from −30°C to +50°C, relative humidity is always in the range of 0–100 (since it's a percentage), and pressure floats around the 100,000 Pascal at sea level.

For certain classifiers, the difference in scale may not be relevant, but for many algorithms, it is mandatory that all the values are in the same ballpark. Some models may even require that the values lie in a specific range (e.g., [−1, +1] or [0, +1]).

Besides this operation of *feature scaling*, the feature engineering step for tabular data is devoted to transforming the input data so that the machine learning model can better pick up the underlying patterns. There are many transformations available. The most common are covered in this chapter. You won't use all of them at the same time in a single project. You must use your judgment every time to understand which fits your current dataset best.

The `tinyml4all` Python package implements a few different operators for this task. To leave this introduction as slim and approachable as possible, let's now deal with only the relevant ones for this project. Please refer to Appendix A for a list of more operators and a more in-depth explanation of those introduced here.

Feature Scaling

Feature scaling, sometimes called normalization, is a transformation that maps each feature's domain into a new domain. You may want to perform this operation because the machine learning algorithm requires (or works better with) data on the same scale or to make it easier to compare different features that don't share the same scale (as in the case of the weather station).

Different forms of feature scaling can be grouped into two categories based on how the scaling factor is applied.

- **Instance-based scaling** only uses the data from the current sample to implement the mapping.

- **Population-based scaling** first scans the entire dataset to compute a set of statistics (e.g., min, max, mean, standard deviation), then uses these statistics to scale each observation.

The most widespread feature scaling strategy is probably the *min-max normalization*. As the name suggests, this normalization first computes the minimum and maximum of each column of the tabular dataset. Then, it maps each sample's values to the range [0, 1] according to the following min-max normalization formula.

$$x \frac{x - min(x)}{max(x) - min(x)}$$

CHAPTER 2   TABULAR DATA CLASSIFICATION

Listing 2-15 shows how to apply min-max normalization to a tabular dataset using the tinyml4all package.

Listing 2-15. Apply Min-Max Normalization to Table

```
from tinyml4all.tabular.features import Scale

minmax = Scale(method="minmax")
table 2 = minmax(table)
```

	r	g	b
count	150	150	150
mean	0,4019	0,2333	0,2344
std	0,231	0,1098	0,1016
min	0	0	0
25%	0,1887	0,1618	0,1842
50%	0,3962	0,2157	0,2368
75%	0,5755	0,2941	0,2895
max	1	1	1

Inspecting the output, you can see that every column now falls in the [0, 1] range (min and max rows), regardless of their original domains.

One of the downsides of min-max normalization is that it is *sensitive to outliers*. Consider the following list of values, where 1000 is the outlier.

0, 20, 21, 22, 23, 24, 25, **1000**

After min-max normalization, the list becomes

0, 0.02, 0.021, 0.022, 0.023, 0.024, 0.025, **1**

Even if most of the data is floating around the value 20, the presence of an outlier determines a "compression" of the range of interest. Without the outlier, the normalized series would have been as follows.

0, 0.8, 0.84, 0.88, 0.92, 0.96, 1

It has a much higher spread factor. So, before you apply min-max normalization, beware of outliers.

Caution   Min-max normalization is sensitive to outliers!

How to Identify Outliers

How do you identify outliers? It is as easy as looking at the table summary. Look at how the values change for each column, moving from min to max. If the difference between each two consecutive rows is in the same figure, then your data is evenly distributed. But if the difference between min and 25% or 75% and max is much larger than the average, your column probably contains outliers.

```
        column_with_outliers
count           106.000000
mean             -0.249395
std               1.782681
min              -8.000000
25%              -0.972149   # 25% - min = 7 ← sign of
                             outliers
50%              -0.364116   # 50% - 25% = 0.6
75%               0.637160   # 75% - 50% = 1
max               7.000000   # max - 75% = 6.4 ← sign of outl
```

For the best feature scaling to apply in the presence of outliers, refer to *robust scaling* in Appendix A.

CHAPTER 2   TABULAR DATA CLASSIFICATION

CHAPTER 2   TABULAR DATA CLASSIFICATION

CHAPTER 2 TABULAR DATA CLASSIFICATION

Figure 2-11 summarizes in a plot the results of the feature scaling methods enumerated so far (plus those from Appendix A).

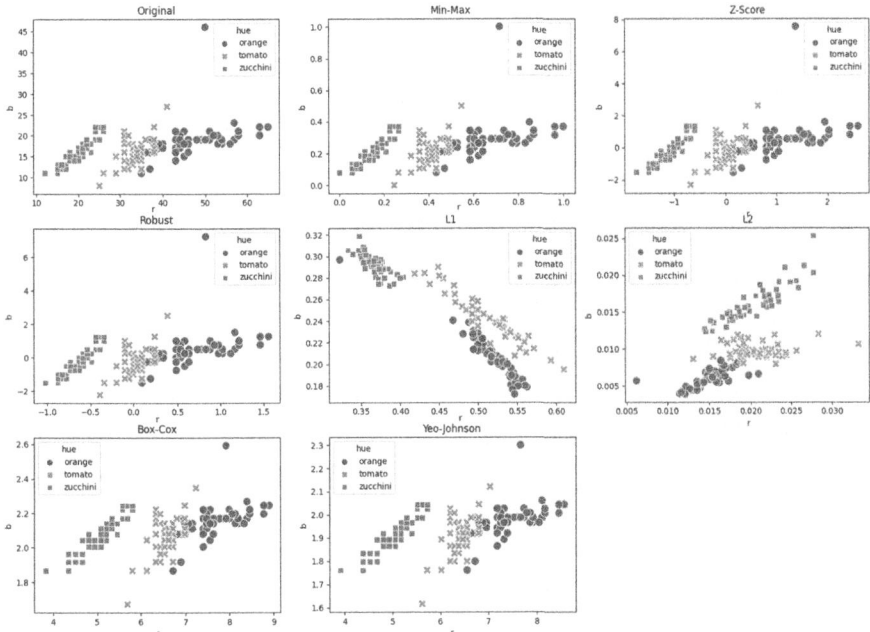

Figure 2-11. *Comparison of feature scaling methods*

It is interesting to note that min-max, z-score, and robust scaling do not alter the distribution of the data (since they're population-based, linear methods).

CHAPTER 2   TABULAR DATA CLASSIFICATION

Feature Selection

Chapter 1 introduced the notion of feature selection. Let's recap its goals.

- Prevent the curse of dimensionality
- Drop features that are highly correlated (redundant)
- Drop features that do not characterize a class (irrelevant)
- Achieve the highest possible metric (e.g., accuracy, F1 score, RMSE) with the lowest number of features

The optimal way to perform feature selection would be to start with all the features and iteratively remove one column at a time. If the classification metrics stay almost the same, then that feature is not very important. Repeating this process until there's no way to keep the metric as-is would yield the best result possible.

If you have few features and a small dataset, this path may be doable. But this process can often take prohibitively long to complete, and you may be interested in quicker, more approximate alternatives.

Since this project only deals with three input variables, feature selection would be of little use here. Since it comes in handy in later chapters, let's quickly enumerate the available options.

Sequential Feature Selection

This is an exhaustive approach. It tries all the combinations of columns to finally achieve the optimal result. Depending on the ratio of features that you want to keep, it may be faster to either drop features one at a time or instead start with an empty list and add one feature at a time (choosing the one that most improves the metric). This is specified by the direction parameter. With this specific method, you can manually choose the number of resulting features to keep or rely on the autoselection mechanism to find the best number for you (see Listing 2-16).

Listing 2-16. Apply Sequential Feature Selection

```
from tinyml4all.tabular.features import Select

# direction = forward will grow from 0 to n features
# direction = backward will shrink from N to n features
select = Select(sequential="auto", direction="forward")
table2 = select(table)
print(table2.head())
     r
0   38
1   40
2   49
3   65
4   54
```

Using the autoselection mechanism, it was found that the Red column alone is sufficient to achieve a good classification result.

Score-based Selection

A faster, classifier-free method to decide which features to drop is based on univariate statistical tests. These tests produce a score for each column; the score represents how important that feature *looks* for classification tasks. I used the term *looks* because you can't be sure how that variable influences the specific classifier that you will use later, so you rely on a statistical approach. In this case, you must manually choose the number of columns to keep. There's no auto setting (see Listing 2-17).

Listing 2-17. Apply Univariate Feature Selection

```
# 1 is the number of columns to select
select = Select(univariate=1)
table2 = select(table)
print(table2.head())
     r
0   38
1   40
2   49
3   65
4   54
```

Also, the univariate test selected the Red column. However, this may not be the case for every dataset.

Recursive Feature Elimination (RFE)

RFE can be looked at as a faster version of sequential feature selection. Again, it requires a classifier to decide which features to drop recursively. The classifier must satisfy a requirement: it must produce a feature importance score (not all classifiers do). This score is used to select the column that contributed the least to the classification result. This process is repeated recursively until the desired number of features is achieved Listing 2-18).

CHAPTER 2   TABULAR DATA CLASSIFICATION

Listing 2-18. Apply Recursive Feature Elimination

```
# 1 is the number of columns to select
select = Select(rfe=1)
table2 = select(table)
print(table2.head())
    r
0  38
1  40
2  49
3  65
4  54
```

Table 2-2 lists a few guidelines that you can use to choose which one of the described methods to use in your specific project.

Table 2-2. *Guidelines to Choose the Best Feature Selection Method*

	Pros	Cons
Univariate	Fast; recommended in the early stages of development	Not guaranteed to result in the optimal selection
Recursive	Faster than sequential; likely converges to the optimal solution; prefer to univariate on medium datasets	May become slow with certain classifiers on large datasets
Sequential	Likely converges to the optimal solution	May become slow with certain classifiers on large datasets

Note You can also manually choose the columns to keep with Select(include=["r", "b"]) or to discard with Select(exclude=["g"]).

81

Back to our fruit classification project: let's consider which feature engineering methods look the most appropriate.

- There are only three features, so feature selection brings little value.
- All the features are continuous, and discretization doesn't make much sense. (We could optionally apply binning, but it would be of little value in this case).
- Feature scaling can benefit some classifiers (and there aren't any outliers).

Given these highlights, let's only implement min-max normalization for now (see Listing 2-19).

Listing 2-19. Apply Feature Engineering to Our Fruits Dataset

```
from tinyml4all.tabular.classification import Table, Chain
from tinyml4all.tabular.features import Scale

table = Table.read_csv("fruits.csv")
# use a Chain to accommodate the case for more steps
optimal_chain = Chain(Scale(method="minmax"))
table2 = optimal_chain(table)
```

Classification Models

The next step of our tiny machine learning workflow revolves around classification models. This step trains an algorithm to recognize the distinctive characteristics of each fruit.

If you did a good job during the feature engineering step, this part should proceed smoothly. You should get acceptable results out of the box, and your focus is optimizing (a.k.a. tuning) the model parameters you chose to increase the accuracy/F1 score to its maximum. Some models may require more work than others to exhibit good results. Next, let's discuss the classifiers available in the tinyml4all package, highlighting the ones you should use until you master the pros and cons of each to make an informed choice.

The following are the main distinctive traits that characterize each type of classifier.

- **Robustness/invariance to feature scaling**: Some classifiers (e.g., decision tree) can classify data no matter the input scale (so normalization can be skipped). Others (e.g., logistic regression) work better if features share a common scale.

- **Linearity/non-linearity**: Some classifiers only work with linearly separable data (e.g., linear SVM), while others allow for polynomial or radial boundaries (e.g., SVM with Gaussian kernel). Some others are rule-based (e.g., decision tree).

- **Memory requirements**: Since our final goal is to deploy the trained model to a resource-constrained device, you should always remember that we want the model to be as small as possible and run as quickly as possible. Decision tree-based classifiers run really fast and require almost zero RAM (they compile to *if-then-else* blocks). In contrast, others perform matrix multiplications to compute the result, meaning they have to store some weights in memory. Their memory, therefore, increases linearly (or even more) with the number of input features.

All the classifiers in the tinyml4all package share the same interface, so you can swap one for another without changing the rest of the code.

CHAPTER 2 TABULAR DATA CLASSIFICATION

Decision Tree

A decision tree (see Figure 2-12) is probably the most intuitive classifier. As its name suggests, it creates a tree data structure that progressively splits on the dataset features to come up with a decision.

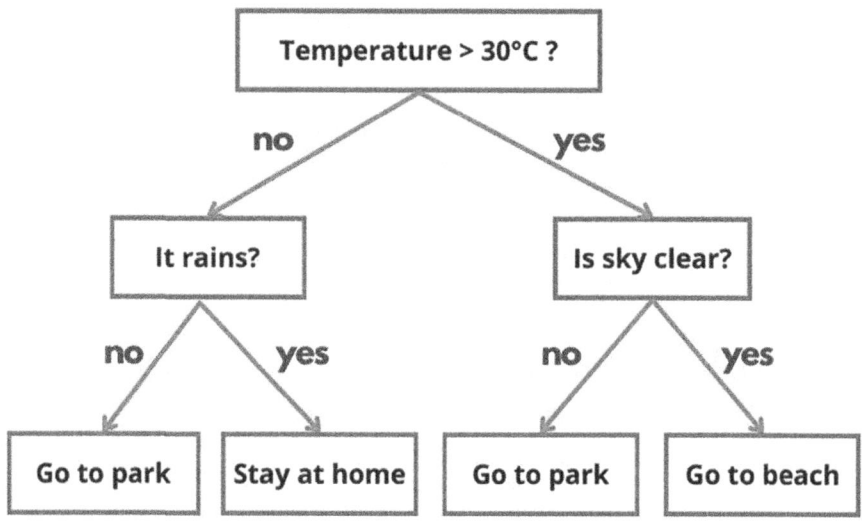

Figure 2-12. Example of decision tree classifier

At each step, the algorithm has to choose which feature to split and on which value. How is this choice made? It is done based on the information gain principle (you can find a gentle introduction to this concept at [4]). In plain words, a split is made where the separation among the different classes improves the most: the more a class becomes clustered after a split, the more the information content increases.

Pros

Decision trees are commonly used in many tasks because they exhibit a few desirable properties.

- **Simple to understand and interpret**: You can look at a tree and clearly understand why it made a given classification.

- **It requires little data preparation**: You don't need normalization, and it works out of the box with non-numerical (categorical) data. You may only want to evaluate if binning improves the results.

- **Innate feature selection**: Unimportant features are automatically ignored since they won't produce any significant information gain.

- **Very fast inference time**: Once on your microcontroller, the tree translates to a plain list of if-then-else without *any* mathematical computation.

Cons

Of course, a decision tree is not a good fit for all projects since it still suffers from the following disadvantages.

- **Tendency to overfitting**: The learned model can be overly complex and fitted to the training data. This can be easily alleviated by parameter tuning (for example, limiting the depth of the tree).

- **Instability**: Small variations in the data might result in a completely different tree being generated.

- **Optimality not guaranteed**: Since practical implementations are based on some heuristics to speed up the learning process (which would otherwise take too long), it can happen that the resulting tree is not the optimal one.

- **Some concepts are hard to learn**: This is because decision trees do not express them easily, such as XOR, parity, or multiplexer problems.

- **Bias on unbalanced datasets**: Decision tree learners create biased trees if some class dominates. It is recommended to balance the dataset prior to fitting with the decision tree.

Random Forest

Simply put, *random forest* [5] is an ensemble of decision trees. Instead of training a single decision tree, random forest trains many different trees, each on a different subsample of the training set: this greatly reduces the chances of overfitting and generally boosts the ensemble accuracy.

Even more than selecting a subsample of the training dataset in terms of observations, random forest applies a technique called *feature bagging*, which feeds to each tree only a random subset of features. This prevents even more overfitting. So, given an input dataset of n observations of m attributes, each tree receives as input n' observations made of m' attributes, with n'< n and m'< m.

At inference time, each tree predicts its own output class; the class with the most votes wins and is selected as the ensemble prediction.

Pros

- The same as decision tree pros
- It has very good accuracy out of the box: random forest is one of the most straightforward classifiers to use since it performs very well with little to no tuning. Setting a reasonable default for the number of trees (10–20) suffice. You can always tune the inner decision tree parameters for top-notch accuracy (max depth, split logic, etc.).

Cons

- The same as decision tree cons
- Black box—in contrast to decision tree, random forest loses its explainability because you can no longer grasp why it predicted a given output due to combining a multitude of trees.

Tip Thanks to its good accuracy out-of-the-box and insensitivity to many feature engineering pre-preprocessing, I recommend you use random forest as your default classifier.

Extreme Gradient Boosting

Extreme gradient boosting (XGBoost) [6] has gained popularity since its introduction in 2014 because it has won many online machine learning competitions. It is, first of all, a *gradient boosting* technique: similarly to random forest, it combines multiple decision trees. Differently from random forests—which trains the trees *in parallel and* where each tree is independent of the others—Gradient Boosting trains them *in series*. Each tree is trained on the results of the previous one: at each iteration, the algorithm tries to reduce a loss function (how badly it performed on the training data).

How? Learning from the errors of the previous one. The core idea of XGBoost is to model the residual errors in cascade to eventually correct them. The "extreme" in Extreme Gradient Boosting emphasizes several software optimizations that greatly improve training time and prediction accuracy.

Pros

- The same as random forest pros
- Very robust to overfitting

Cons

- The same as random forest pros
- Slower than random forest

Logistic Regression

Despite its name, logistic regression is used as a binary classification algorithm. It borrows its name from the linear regression, which produces a linear model in the form described in the following linear regression output formula.

$$y_i w^T \cdot x_i$$

x_i is the input sample, and w is a weight vector. This means that each value in the input sample is multiplied by a weight, and the result is accumulated. This produces a continuous output (as in the regression).

To move from regression to classification, logistic regression applies the logistic function to the output so that its domain becomes [0, 1]: in the binary classification case, this value represents the probability that the current sample belongs to the positive class. Figure 2-13 displays the logistic function plot.

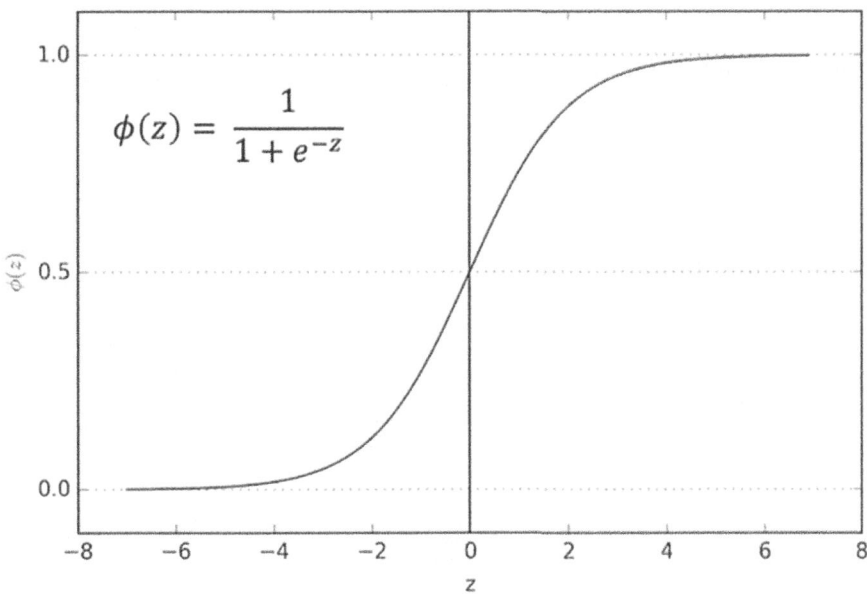

Figure 2-13. *The logit function*

The objective of the training phase for logistic regression is to minimize the classification error or, conversely, to maximize the likelihood.

The following computes the probability of sample i-th belonging to the positive class.

$$p_i = logit(w^T \cdot x_i)$$

The likelihood of the entire dataset is computed in the following formula.

$$L(w) = \sum_{i=1}^{N} y_i ln(p_i) + (1-y_i) ln(1-p_i)$$

CHAPTER 2 TABULAR DATA CLASSIFICATION

y_i is the ground truth label of the sample (either 0 or 1 since this is a binary classifier). You can interpret the formula using the following intuition:

- If y_i is 1 and $p_i > 0.5$, the likelihood increases (correct classification).
- If y_i is 0 and $p_i < 0.5$, the likelihood increases (correct classification).
- If y_i is 1 and $p_i < 0.5$, the likelihood decreases (wrong classification).
- If y_i is 0 and $p_i > 0.5$, the likelihood decreases (wrong classification).

At the end of the training, the weight vector w is such that the global likelihood is at its maximum, or equally, we got the highest number of correct classifications. The multiclass case follows a very similar pattern, but its demonstration is out of the scope of this book.

Pros

- Fast inference time (linear in the number of features)
- Predicts a probability of class membership
- The weight coefficients could indicate the "importance" of a feature for the class membership

Cons

- Sensitive to noise and outliers
- The number of weights grows linearly with the number of inputs
- Linear model, so you'd better transform your input data to improve the accuracy (apply power, log, square root operations)

- Sensitive to correlated input
- May not converge to an optimal value for *w*

Support Vector Machines

Support vector machines (SVM) are a binary classification algorithm applicable to multiclass problems using the one vs. one voting scheme (refer to Chapter 1).

The core idea of SVM is to find a separation boundary between the two classes with a margin as large as possible to provide a robust classification. Figure 2-14 is a visual intuition.

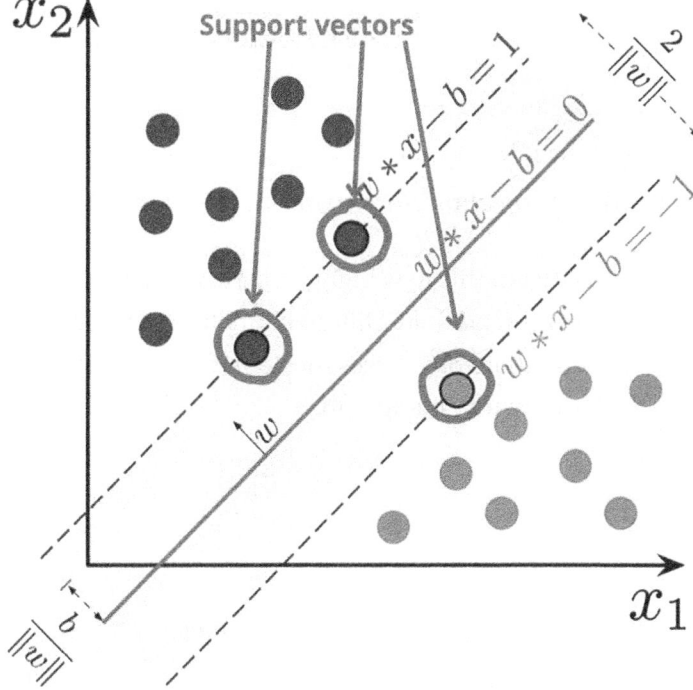

Figure 2-14. *SVM separation margin*

The margin is computed only from a few meaningful points in the training set: the *support vectors*. These points are the most difficult to classify since they live on the border between the two classes.

However, in real-world data, it is often not easy to find a clear separation between the two classes—often, this separation doesn't even exist! To solve this problem, the main intuition of SVM is to project the original data in a higher-dimensional domain. If the original data is 1D, for example, it can be projected into a 2D space (see Figure 2-15).

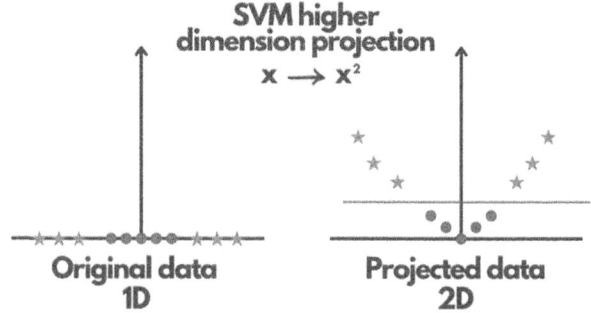

Figure 2-15. SVM projection into higher-dimensional space

The rationale is that in this new highly dimensional space, the two classes are more clearly separable. Different functions can perform such projection, and the math involved uses the *kernel trick*, but that falls out of the scope of this introductory paragraph.

Pros

- Effective with highly dimensional data
- Only uses the support vectors for inference
- Robust against outliers

Cons

- Training time doesn't scale well with the number of training samples
- Doesn't perform very well in the presence of many overlapping classes
- Requires tuning of multiple parameters to perform well
- Works only on numeric inputs
- Black box model
- With n features and k support vectors, the number of weights to store in RAM is n * k (k being larger the less separable the classes, at least one per class—often more)

Table 2-3 summarizes the advantages and disadvantages of classier types.

Table 2-3. Comparison of Classifiers

	Pros	**Cons**
Decision Tree	Fast to train and execute; insensitive to features' scale; memory-efficient once deployed	Prone to overfitting and bias in unbalanced settings; may exhibit suboptimal accuracy
Random Forest	The same as decision tree pros; shows better accuracy out-of-the-box	Not interpretable

(*continued*)

Table 2-3. (*continued*)

	Pros	Cons
XGBoost	The same as random forest pros; can add a few % in accuracy	Slower to train than random forest
Logistic Regression	Inference time and memory linear in the number of features	Inference time and memory linear in the number of features; linear model; sensitive to outliers
SVM	Effective with high-dimensional data; robust against outliers	Slow to train on large datasets; requires parameter tuning to work well; inference time and memory more-than-linear in the number of features

Now that the available classifiers have been enumerated, let's look at how to apply them to the fruits project (see Listing 2-20). Classifiers are no different from feature engineering steps in their API.

Listing 2-20. Train Classifier and Make Predictions

```
from tinyml4all.tabular.classification.models import
DecisionTree, RandomForest, LogisticRegression, SVM

# print a description of the parameters you can set
# on the given classifier
print(help(RandomForest))

# train classifier and make predictions
rf = RandomForest(n_estimators=20)
table2 = rf(table)
# table.full() prints the data + true labels + predicted labels
print(table2.full())
```

CHAPTER 2 TABULAR DATA CLASSIFICATION

	r	g	b	truth	__prediction__
0	38	23	18	orange	orange
1	40	22	18	orange	orange
2	49	25	19	orange	orange
3	65	31	22	orange	orange
4	54	25	18	orange	orange

To quickly inspect how well your classifier performed overall without comparing each row manually, you can generate a classification report with the code in Listing 2-21.

Listing 2-21. Generate a Classification Report

```
print(table2.classification_report())
              precision    recall  f1-score   support

      orange       1.00      0.90      0.95        50
      tomato       0.89      1.00      0.94        50
    zucchini       1.00      0.98      0.99        50

    accuracy                           0.96       150
   macro avg       0.96      0.96      0.96       150
weighted avg       0.96      0.96      0.96       150

+--------------------+--------+--------+----------+
| True vs. Predicted | orange | tomato | zucchini |
+--------------------+--------+--------+----------+
|             orange |   45   |   5    |    0     |
|             tomato |   0    |   50   |    0     |
|           zucchini |   0    |   1    |    49    |
+--------------------+--------+--------+----------+
```

The report contains a detailed list of metrics for each class (precision, recall, F1 score), overall accuracy, and the confusion matrix. You can leverage this information to judge your classifier's performance and compare different classifiers' results.

Classification Chain

So far, you've seen how to perform feature engineering and classification in Python by considering each component on its own. But these steps don't live in isolation: you must always perform them in the same order to get reproducible results. You can do this manually if you want, but to make this process more manageable and to be able to export the workflow to C++ later, you need to create a *chain*.

A chain is, at its core, a list of operations that act sequentially on the data. Each operation produces an output fed as input to the next until the final result is generated. The Chain class can handle any feature scaling, feature selection, and classification operator described so far. Listing 2-22 shows how to instantiate a full classification chain for the fruit dataset.

Listing 2-22. Create a Classification Pipeline for a Table Object

```
from tinyml4all.tabular.classification import Table, Chain
from tinyml4all.tabular.classification.models import
RandomForest
from tinyml4all.tabular.features import Scale

table = Table.read_csv("fruits.csv")
chain = Chain(
    Scale("minmax"),
    RandomForest()
)
classified = chain(table)
```

CHAPTER 2   TABULAR DATA CLASSIFICATION

	r	g	b	truth	prediction
0	0,491	0,294	0,263	orange	orange
1	0,528	0,275	0,263	orange	orange
2	0,698	0,333	0,289	orange	orange
3	1	0,451	0,368	orange	orange
4	0,792	0,333	0,263	orange	orange

The first time you call a chain on a table, it goes through each step and fits it (to learn its internal parameters). After the initial training phase, you can apply it to any other table: this time, the chain remembers its parameters. It only applies the corresponding transformation to the new table (see Listing 2-23).

Listing 2-23. Apply Classification Pipeline to a New Dataset

```
table1 = Table.read_csv("dataset_1.csv")
table2 = Table.read_csv("dataset_2.csv")
chain = Chain(
     Scale("minmax"),
     RandomForest()
)

# train chain on table1
chain(table1)

# apply chain to table2
# this will use the parameters learned on table1
# e.g., use min/max from table1
table2_classified = chain(table2)
```

97

CHAPTER 2 TABULAR DATA CLASSIFICATION

Deploy to Arduino

At this point, the machine learning workflow is done. We implemented feature engineering and classification on our powerful desktop PC; the accuracy achieved by the classifier meets our expectations.

Here is where the *tiny* part comes in. The last step is to finally convert our pipeline to C++ so that we can import the code into our embedded project.

Listing 2-24 shows how to do this and save the result to a file called FruitChain.h in the same folder as your Python script.

Listing 2-24. Convert Pipeline from Python to C++

```
chain.convert_to("c++", class_name="FruitChain", save_to="FruitChain.h")
```

The generated code is a stand-alone class that you can import into any C++ project (Arduino, PlatformIO, ESP-IDF, etc.). You don't really have to dig into how it works internally: you only need a single line to invoke it. This code works equally well on almost any microcontroller that supports C++ (8- or 32-bit, with or without Floating Point Unit, from ARM, Atmel, or Espressif).

How to Use in Arduino Sketch

We now have all the tools needed to complete our fruit classification project. The code to configure and read the color sensor is the same as Listing 2-2. Listing 2-25 adds the logic to classify the fruit and print the result to the Serial Monitor.

Listing 2-25. Arduino Sketch to Classify Fruit Based on Color

```
/**
 * Listing 2-25
 * Predict fruit from RGB color components.
 *
 * Required hardware: Arduino Nano BLE Sense.
 */
#include <Arduino_APDS9960.h>
#include <tinyml4all.h>
// put the generated file from Python inside the Arduino
// sketch folder!
#include "./FruitChain.h"

tinyml4all::APDS9960 sensor;
tinyml4all::FruitChain chain;

void setup() {
    Serial.begin(115200);
    while (!Serial);
    Serial.println("Fruits classification example");

    sensor.begin();
}

void loop() {
    sensor.readColor();

    // chain(input) will return true on success
    // false on error
    if (!chain(sensor.r, sensor.g, sensor.b))
      return;

    // the predicted human-readable label is in
    // chain.output.classification.label
```

CHAPTER 2 TABULAR DATA CLASSIFICATION

```
    // the numeric output (0, 1, 2, ...) is in
    // chain.output.classification.idx
    Serial.print("I think this is ");
    Serial.println(chain.output.classification.label);

    delay(1000);
}
```

Upload the sketch and open the Serial Monitor. You see that the board makes a classification each second. Put the different fruits in front of the sensor and watch the results update accordingly (see Figure 2-16).

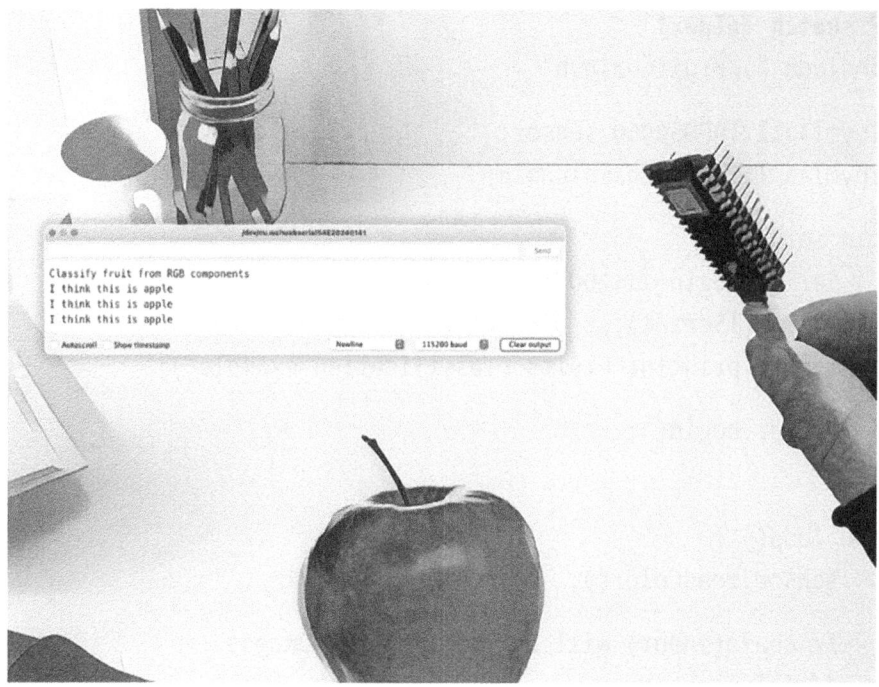

Figure 2-16. *Live classification of fruit from RGB components*

Congratulations, you completed your first TinyML project!

CHAPTER 2 TABULAR DATA CLASSIFICATION

Warnings

I want to draw your attention to a detail that you may have missed. Point your color sensor at something different from your fruits—the wall or your desk. What does the classifier predict?

It may surprise you that the classifier is still predicting one of the fruits. You could expect that the classifier would have detected that it was not pointing to any of the fruits and responded accordingly. Sadly, it doesn't work this way. Recall that in a classification task, the list of classes is completely known, meaning that a model cannot detect a class it has never seen before.

We trained our model only on the fruit classes. It doesn't have the concept of a "wall" or a "desk," so it will never predict that. If we wanted to teach the classifier to recognize that it is not looking at any fruit, we must also show the data of these cases.

Caution Classifiers only recognize cases they have seen during training. If you have a case for "no object of interest" (e.g., no fruit), you need to collect data for it.

Summary

This chapter was dense with new information. You learned how to collect data from your microcontroller either manually or with the help of automated tools; then you got an overview of the feature engineering methods available to you to make the raw data you collected easier to later classify and (a subset of) the many models that make the machine learning landscape. As a last step, you were able to convert the tools required to perform an end-to-end classification pipeline from Python to performant C++ so that they can be imported into any embedded project.

The next chapter also focuses on tabular data. Given classification was covered in this chapter, the next chapter expands on regression.

CHAPTER 3

Tabular Data Regression

Tabular data regression shares many of the steps required to perform tabular data classification (namely, data capture and feature engineering). It differs for the data plotting and the algorithm selection part. This chapter features a project that highlights the workflow required for a regression task but moves quickly through the topics covered in Chapter 2. The primary focus of this chapter is to showcase the peculiarities of regression.

The project that guides us through this chapter is a *proximity meter* inferred from RGB color components. Chapter 2 demonstrated that color components can be used to classify colored objects. However, color components depend on ambient light (the greater the illumination, the higher the values). You can leverage this phenomenon to *estimate* how far the sensor is from a flat surface: the more you approach the surface, the more shadow the board and your own body cast on it, thus lowering the detected intensities. The task is to model this relationship between color intensities and distance (see Figure 3-1).

CHAPTER 3   TABULAR DATA REGRESSION

Figure 3-1. The result of distance prediction from RGB components

The steps required to complete the project are the same as those outlined in Chapter 2.

1. Collect data (RGB + distance in this case).
2. Load and inspect data to spot anomalies and bad data and understand the relation between input and output (if it exists!).
3. Perform feature engineering to boost the modeling efficacy.
4. Train a machine learning model for the regression task and assess its performance.
5. Convert the entire regression chain to C++ and deploy the code to our board.

Since many aspects are identical to what you've seen working with tabular data, this chapter focuses on those who change from classification to regression.

Required Hardware

As in Chapter 2, you need a color sensor to reuse the same hardware you already have. In addition to that, you need a distance sensor. Since this toy project does not need to achieve millimeter accuracy, any distance sensor will do the job (ultrasonic, time-of-flight, IR, etc.).

The Arduino Nano BLE Sense board has a proximity sensor built-in (the same APDS that detects the colors), but I found that it doesn't work well at low distances (<5 cm), so I preferred to hook an external ultrasonic sensor (see Figure 3-2).

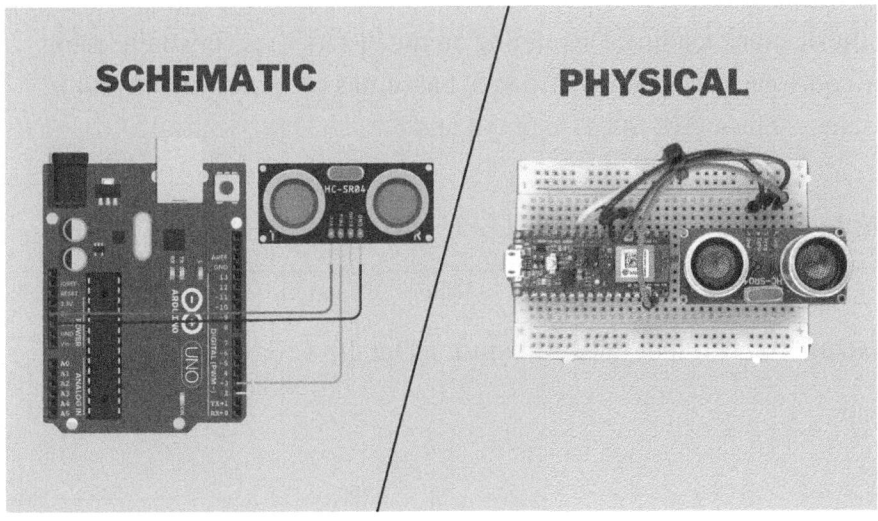

Figure 3-2. Ultrasonic circuit diagram

CHAPTER 3   TABULAR DATA REGRESSION

The ultrasonic distance sensor uses *sound waves* to measure the distance to an object. It follows the *time-of-flight* principle, measuring the time an ultrasonic pulse takes to travel to an object and back. When the TRIG pin is set to high, the sensor emits an ultrasonic sound wave at a frequency of 40 kHz via its transmitter; the emitted sound wave travels through the air and reflects off an object in its path. The sensor's receiver detects the reflected wave and outputs a high signal on the ECHO pin to the Arduino. The measured time between the trigger and the echo is directly proportional to the distance of the object via the speed of sound in the air (approximately 343 meters per second).

Capture Data

Let's leverage the Python collection script from Chapter 2 to capture data. The Arduino sketch is similar to the color collection one, with the addition of the distance reading. Depending on the distance sensor you're using, the code may vary slightly. Listing 3-1 assumes you wired an ultrasonic distance sensor (HC-SR04) to pins 4 and 5.

Note   Replace the TRIG and ECHO pins with your own!

Listing 3-1. Collect RGB + Distance Data in CSV Format

Nano
```
/**
 * Listing 3-1
 * Collect RGB + distance data in CSV format (unattended).
 *
 * Required hardware: Arduino Nano BLE Sense.
 * Required hardware: Ultrasonic distance sensor (HC-SR04)
```

```
  */
#include <Arduino_APDS9960.h>
#include <tinyml4all.h>

#define ECHO 2
#define TRIG 3

using tinyml4all::printCSV;

tinyml4all::APDS9960 sensor;
tinyml4all::Ultrasonic ultrasonic(ECHO, TRIG);

void setup() {
  Serial.begin(115200);
  while (!Serial);
  Serial.println("Collect RGB + distance as CSV");

  // init color and distance sensors
  sensor.begin();
  ultrasonic.begin();
}

void loop() {
  // read R, G, B
  sensor.readColor();

  // read distance in millimeters
  uint16_t distance = ultrasonic.millimiters();

  // print data as CSV
  printCSV(sensor.r, sensor.g, sensor.b, distance);
  delay(1000);
}
```

Flash the sketch to your board and confirm it works by opening the Serial Monitor (see Figure 3-3).

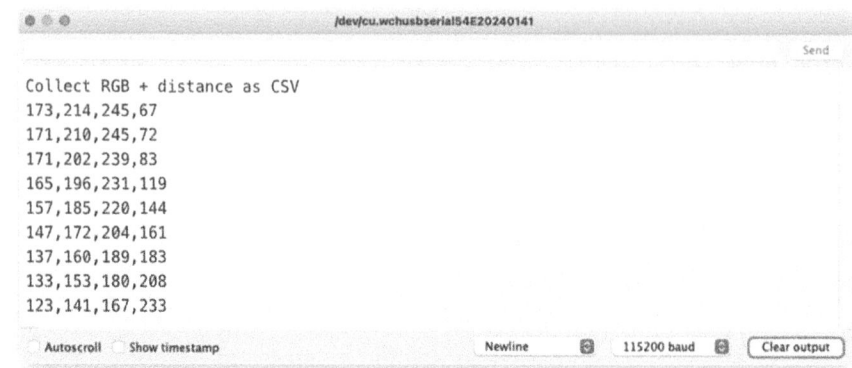

Figure 3-3. *RGB distance collect Serial output*

The Python script shown in Listing 3-2 is a replica of Listing 2-3 from Chapter 2, reported here for convenience. The only modification is that we now continuously poll data from the board without asking for fruit names and the number of samples.

Listing 3-2. Capture Data from Serial in Python

```
from tinyml4all.tabular import capture_serial

capture_serial(
    # * is a wildcard character that matches anything
    # on Windows, this will look like COM1 or similar
    port="/dev/cu.usb*",
    baudrate=115200,
    # file where data will be saved
    save_to="rgb-distance.csv",
    # name of columns
    headings=""r,g,b,distance"",,
    num_samples=100
)
```

CHAPTER 3   TABULAR DATA REGRESSION

To run the code, open a terminal inside the folder where the script is located and run.

```
(tinyml)$ python capture_rgb_distance.py
Press [Enter] when you're ready to start:
Task will start in 3...2...1...START!
100%|██████████████████████| 100/100 [00:100<00:00, 1s/it]
Collected 100 lines of data
```

During this phase, start from 2 to 3 cm away from a bright, flat surface and slowly move the board/sensor far and far away (consider that it is taking one reading per second, so do it in small steps).

When the process finishes, be sure you have a rgb-distance.csv file with 100 lines of data, then move on.

Load and Inspect Data

The data captured is still in tabular form, so you can reuse all the listings from Chapter 2 to manipulate the table instance. In this case, there is a single file with all the data, so the relevant code to load it is replicated in Listing 3-3.

Listing 3-3. Load a Single CSV File

```
# note that this time the module is called regression
# instead of classification!
from tinyml4all.tabular.regression import Table

table = Table.read_csv("rgb-distance.csv")
table.set_targets(column="distance")
print(table.describe())
```

CHAPTER 3   TABULAR DATA REGRESSION

	r	g	b	__target__
count	100	100	100	100
mean	133,9	164,66	189,44	244,64
std	28,664	38,523	40,609	132,33
min	80	94	110	49
25%	108	128,75	153	141,5
50%	143,5	178	204	219,5
75%	160	201,25	226,75	354
max	168	210	232	480

The code to load a table for regression is the same as for classification, and the responsible class is still called Table. Pay attention, though, to the import statement: it changed from tinyml4all.tabular.classification to tinyml4all.tabular.regression! The set_targets function didn't change either in its form. But the underlying logic is a lot different: while for classification tasks, the targets are *labels,* now they represent continuous values—and they're treated like the rest of the columns in the table summary report.

Once data is loaded, you can plot it.

Plot Regression Data

In the context of regression, colored scatter plots and pair plots, as you saw in the previous chapter, have no meaning. They are meant to visualize clusters of data based on class membership. However, regression has no such thing as classes; instead, there are continuous outputs.

Plotting regression data is a bit harder than plotting classification data in the case of many input variables, but you can still deduct a good level of information.

CHAPTER 3   TABULAR DATA REGRESSION

One Input

If your problem has only one input variable, it is straightforward. You plot the input variable vs. the output variable on a Cartesian plane using the code in Listing 3-4. The output is displayed in Figure 3-4.

Listing 3-4. Draw a Scatter Plot of One Input

```
# only plot a single column
table.scatter(column="r")
```

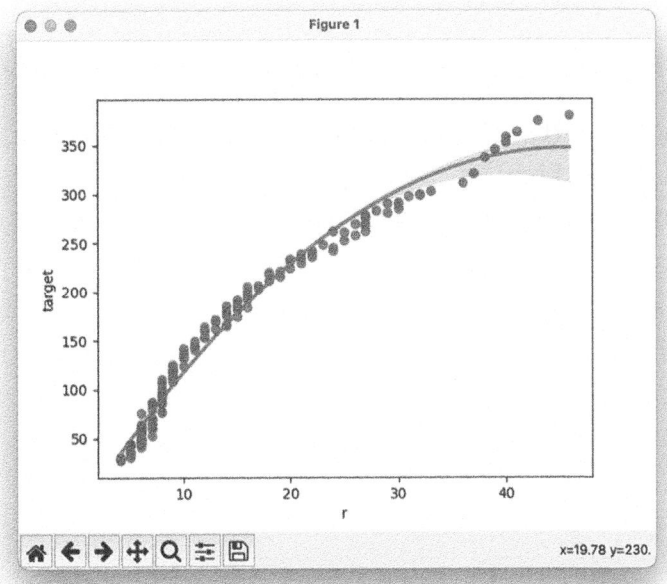

Figure 3-4. *Scatter regression plot of a single column*

The red line you see in the middle is the linear regression output. How that line is defined is demonstrated later; for now, just consider that one the best possible fit for a straight line to match the input data.

111

CHAPTER 3   TABULAR DATA REGRESSION

Many Inputs, Many Scatters

If your data has many inputs, the trivial extension is to plot the target variable against all the inputs (see Listing 3-5 and Figure 3-5).

> **Note**   While it *could* be possible to apply dimensionality reduction to collapse the columns to two or even one variable (like you did for classification), the result would be disappointing. There's a good chance that the resulting variable(s) doesn't exhibit any straight relation with the output variable, so your plot would be misleading.

Listing 3-5. Draw Scatter Plot of Many Inputs

```
# you can stack the plots horizontally or vertically
# default is vertically
table.scatter(orientation="vertical")
```

CHAPTER 3 TABULAR DATA REGRESSION

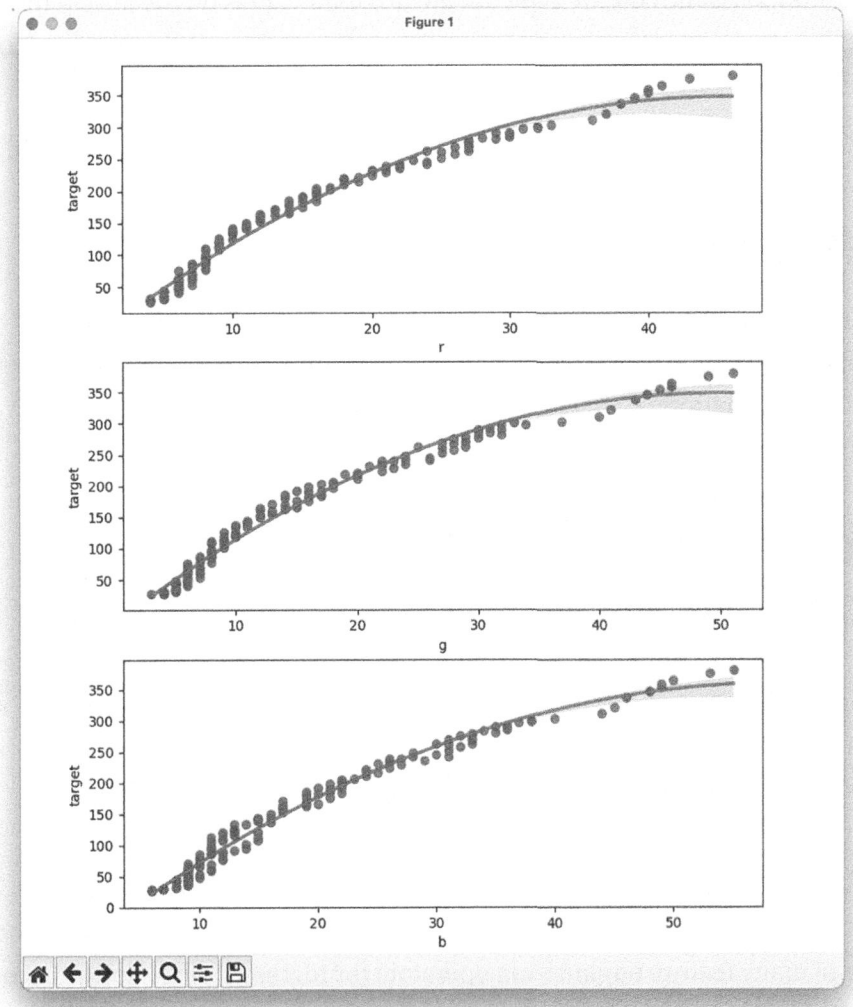

Figure 3-5. *Scatter regression plots of all columns of the table*

By looking at the plots, you can see that there is a strong relation between the color components and the distance since all dots are not too distant from the red line, and they seem to follow a monotonically growing distribution (values on the right are greater than values on the left). If the dots had looked much sparser and more scattered, it would've been a sign that—maybe—a relation between input and output could not be found—at least not without feature engineering first.

Feature Engineering

Many regression models work well with linear relationships between inputs and output(s). Not all data, though, exhibits such a relationship *as-is*. It may happen that a linear relation still exists, but on monotonic functional mappings of the input or a combination of inputs.

Monotonic Functional Mappings

What is a *monotonic functional mapping*?

It simply is a mathematical function applied to the input that preserves its order (i.e., if x1 < x2, then f(x1) < f(x2)). This function can be as simple as the power, logarithmic, or exponential operator. Let's consider the simple formula.

$$y = 3x^2 + 5$$

As simple as it looks, linear regression cannot model this function. The role of many feature engineering operators for regression is to apply one or more mappings so that a new feature is linearly correlated with the output.

$$\begin{cases} z = x^2 \\ y = 3z + 5 \end{cases}$$

At this point, most of the regression models available can predict the outputs with a good degree of accuracy. A wide range of common mappings can be suitable in most cases.

- Power of 2, power of 3, square root, inverse
- Exponential and logarithm
- Box-Cox and Yeo-Johnson

Box-Cox and Yeo-Johnson are discussed in the Appendix A, so refer to it for more information. Listing 3-6 showcases how you can apply the power and exponential/logarithm transforms to a tabular dataset. You can optionally apply only a subset of all the available mappings and apply them only to a subset of columns. The available mappings are (see Figure 3-6).

- `square` (power of 2)
- `cube` (power of 3)
- `sqrt` (square root)
- `inverse` ($1/x$)
- `exp` (exponential – e^x)
- `log` (natural logarithm)

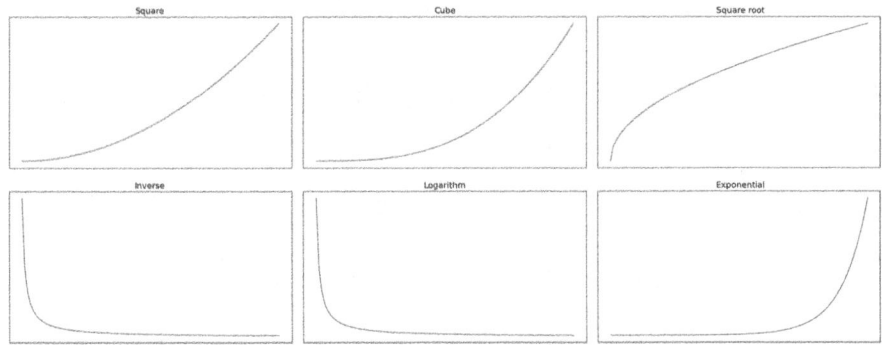

Figure 3-6. *Monotonic functional mapping plots*

Listing 3-6. Apply Monotonic Functional Transforms

```
from tinyml4all.tabular.features import Monotonic

# apply the square and cube mapping to all columns
square_and_cube = Monotonic(functions="square, cube")

# apply all the mappings only to the "r" column
only_r = Monotonic(columns="r")

# apply all mappings to all columns
monotonic = Monotonic()

# run the transform on the table
table2 = monotonic(table)
print(table2.describe())
```

	r	g	b	...	sqrt(g)	inverse(g)	log(g)
count	100	100	100	...	100	100	100
mean	14,36	15,18	19,037	...	3,664	0,105	2,572
std	9,963	11,167	11,442	...	1,328	0,065	0,643
min	4	3	6	...	1,732	0,02	1,386
25%	7	6	10	...	2,449	0,045	1,946
50%	10	11	15	...	3,317	0,091	2,485
75%	20	22	26	...	4,69	0,167	3,135
max	46	51	55	...	7,141	0,333	3,951

Which mappings should you use? It all depends on your data, which is where plotting is handy. By looking at each regression plot, you should be able to associate the *shape* of the dots' distribution to one of the available functions. If that's the case, you can be laser-focused and only apply the

CHAPTER 3 TABULAR DATA REGRESSION

strictly necessary transform to the column that needs it. Otherwise, if you cannot spot any good match, leave it blank so all are tried—and maybe add a feature selection step later.

> **Caution** The exponential operator can grow very quickly to intractable numbers! I strongly recommend you always apply feature scaling before.

Polynomial Input Combinations

Sometimes, your features may carry important information when considered together instead of in isolation.

For example, imagine you want to create a model in the healthcare industry, and you deal with patients' data. It is well known that height and weight, on their own, may not be relevant to determine whether a person is in good shape, but combined, they better describe the status of a person.

The polynomial feature transformation generates new features by multiplying each pair of columns. It also generates *squared* columns (multiplying a column by itself). Listing 3-7 shows how to use it.

Listing 3-7. Apply Polynomial Features Expansion

```
from tinyml4all.tabular.regression import Table
from tinyml4all.tabular.features import Multiply

# assume the people.csv files contains width (w), height (h)
# and BMI of a group of people
table = Table.read_csv("people.csv")
table.set_targets(column="bmi")

# if you omit the columns parameter,
```

```
# all columns will be considered
mult = Multiply(columns=["w", "h"])
table 2 = mult(table)
```

	h	w	h_x_h	w_x_w	h_x_w
0	74	242	5453	58512	17863
1	69	162	4731	26345	11164
2	74	213	5492	45259	15766
3	72	220	5145	48419	15784
4	70	206	4883	42580	14420

Tip Since multiplication can lead to very large values or be skewed if the two columns are in completely different ranges, it is recommended that you first apply feature normalization.

Regression Models

As for classification, regression has many available models from which to choose. They vary by descriptive power and complexity: until you get familiar with them, I suggest you try all for your specific project and select the one that performs best. This section introduces the following models.

- Ordinary Least Squares or Linear Regression
- Ridge
- Lasso
- Decision tree
- Random forest

Ordinary Least Squares

Also known as *linear regression*, the most used model for regression is the ordinary least squares (OLS) method. It assumes that a linear relationship exists between the samples (denoted by X) and the ground truth (denoted by Y). This linear relationship can be expressed in matrix form with the following formula.

$$Y = W^T \cdot X$$

W^T is an unknown weight matrix that we aim to find. Since we often cannot find an exact solution to this equation, this method aims to minimize the sum of squared errors between ground truth and model's predictions.

$$minimize \sum_{i=1}^{N} (truth_i - pred_i)^2$$

This iterative process should converge to the optimal solution after a given number of iterations. Listing 3-7 is an example of how to use this method and the others covered in this chapter.

Ridge and Lasso

One problem with ordinary least squares is that it has no limits on the magnitude of its weights. It may happen, for example, that—for our distance from the RGB project—it comes out with a relation similar to the following.

$$dist = 1432r + 785g + 354b$$

Now, this is a legitimate relationship that *could* exist. But most of the time, in a real-world scenario, we can expect that the magnitudes of the coefficients are much smaller (depending on the scales of the inputs) and that those strange numbers are a sign of *overfitting*. The Ridge [1] regressor tries to minimize the prediction error but also tries to keep the weights as low as possible, thus reducing the chance of overfitting.

The Lasso [2] method is similar in spirit to Ridge in that it also tries to *regularize* the weight parameters. While Ridge's objective is to keep their values as low as possible, Lasso's objective is to set the highest possible number of them to zero, thus creating a *sparse* solution (where only a few of the values are non-zero). This trait can be useful in settings with many (irrelevant) inputs.

Refer to Listing 3-7 for a code example.

Decision Tree and Random Forest

Not all regression relationships are linear (either on the raw input or their transformed versions). The models mentioned earlier may fall short if your data exhibits more complex relationships. One option would be to introduce non-linear features through feature engineering. Another option is to use a highly non-linear model like decision tree (and, by extension, random forest).

These models work the same way in the classification setting: they generate a sequence of splits on the input features. Each split ends in a (continuous) output value, instead of a class label (see Figure 3-7).

CHAPTER 3   TABULAR DATA REGRESSION

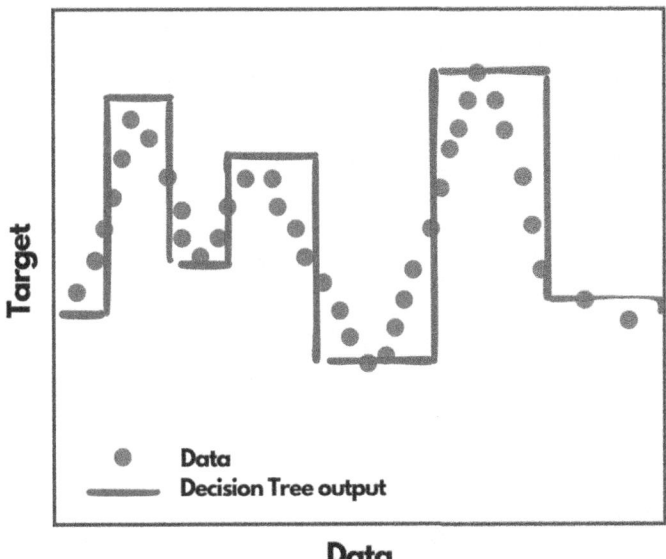

Figure 3-7. Decision tree for regression

Table 3-1 summarizes the characteristics of the models enumerated so far.

Table 3-1. Summary of Regression Models

	Pros	Cons
Ordinary Least Squares (OLS)	Works well with linear data; fast to train; memory linear in the number of features	May overfit; doesn't work well with non-linear data
Ridge	Less prone to overfitting than OLS	Doesn't work well with non-linear data
Lasso	Less prone to overfitting than OLS	Doesn't work well with non-linear data
Decision Tree/ Random Forest	Works well with highly non-linear data	May require parameter tuning not to overfit. Not recommended for very linear data

You can refer to Listing 3-8 to use any of these regressors.

Listing 3-8. Apply Regression Model to Table

```
from tinyml4all.tabular.regression.models import Linear, Ridge,
Lasso, DecisionTree, RandomForest

linear = Linear()
ridge = Ridge()
lasso = Lasso()
tree = DecisionTree()
rf = RandomForest()
# apply any of the models above
table2 = linear(table)

# get metrics
print(table2.regression_report())
+-------+-------+------+------+
|  MAE  |  RMSE | MAPE | R^2  |
+-------+-------+------+------+
| 32.33 | 38.17 | 43%  | 0.80 |
+-------+-------+------+------+
```

Regression Chain

After you have experimented with the different processing operators enumerated so far and inspected both the scatter and the reports, you can finally assemble a complete regression chain that produces the best output possible (Listing 3-9).

For the RGB distance dataset from this project, the following chain, which is made of robust feature scaling, monotonic functional mappings, and linear regression, performed pretty well.

Listing 3-9. Complete Regression Chain for RGB Distance Dataset

```
from tinyml4all.tabular.regression import Table, Chain
from tinyml4all.tabular.regression.models import Linear
from tinyml4all.tabular.features import Scale,
Monotonic, Select

table = Table.read_csv(""rgb-distance.csv"")
table.set_targets(column=""distance"")
chain = Chain(
    Scale("robust"),
    Monotonic(),
    # optional, may decrease accuracy a bit
    #Select(sequential="auto", estimator=Linear()),
    Linear()
)
predictions = chain(table)
print(predictions.regression_report())
+-------+-------+------+------+
| MAE   | RMSE  | MAPE | R^2  |
+-------+-------+------+------+
| 6.29  | 8.19  | 4%   | 1.00 |
+-------+-------+------+------+
```

Since our distance is expressed in millimeters, the results highlight that the average absolute error of our model is 6 mm. This is a reasonable error for our simple project based on a pretty naive correlation between ambient light and distance.

If you're not satisfied, you must revise your chain composition. Since feature scaling improves the results in almost every case, you should focus on the feature engineering part—maybe add a polynomial combination step? The choice of model is whether the relation between input and output is linear. Or would a tree-based model would work better?

CHAPTER 3   TABULAR DATA REGRESSION

Deployment

After you're satisfied with the results of the regression, it is time to port it to C++. A regression chain is no different from a classification chain, as seen in Listing 3-10. The generated code is a stand-alone class that you can import into any C++ project (not necessarily Arduino-based). Copy the generated file inside your Arduino project folder before moving to the next section!

Listing 3-10. Convert Regression Chain to C++

```
chain.convert_to("c++", class_name="DistanceChain", save_to="""
DistanceChain.h""")
```

How to Deployment Use in Arduino Sketch

Similar to Chapter 2, the deployment sketch is almost identical to the capturing sketch: it only adds a few lines to run the regression on the inputs. The chain receives three inputs: the red, green, and blue light components. It then produces a single, continuous output with the estimated distance in millimeters. Listing 3-11 assumes you still have your distance sensor attached so that you can compare the predictions versus the *actual* distances. After you assess the system's field performance, you can remove the distance sensor. (Otherwise, it would be pointless to use an estimated value when you have the measured one!)

Listing 3-11. Arduino Sketch to Predict Distance Based on RGB Components

```
Nano
/**
 * Listing 3-11
 * Predict distance from RGB
```

CHAPTER 3   TABULAR DATA REGRESSION

```
 *
 * Required hardware: Arduino Nano BLE Sense.
 * Optional hardware: Ultrasonic distance sensor (HC-SR04)
 */
#include <Arduino_APDS9960.h>
#include <tinyml4all.h>
// this is the file generated in Listing 3-10
#include "./DistanceChain.h"

// replace with your own pins, if different
#define ECHO 2
#define TRIG 3

using tinyml4all::printCSV;

tinyml4all::APDS9960 sensor;
tinyml4all::Ultrasonic ultrasonic(ECHO, TRIG);
tinyml4all::DistanceChain chain;

void setup() {
  Serial.begin(115200);
  while (!Serial);
  Serial.println("Predict distance from RGB");

  // init sensors
  // will throw an error if something goes wrong
  sensor.begin();
  ultrasonic.begin();
}

void loop() {
  // read R, G, B
  sensor.readColor();
```

125

CHAPTER 3 TABULAR DATA REGRESSION

```
  // feed data to the regression chain
  // should always return true
  if (!chain(sensor.r, sensor.g, sensor.b))
    return;

  // the chain output is in output.regression.value
  Serial.print("Predicted distance: ");
  Serial.print(chain.output.regression.value);
  Serial.print(" mm");

  // if a distance sensor is available,
  // print the error between measured and predicted
  uint16_t distance = ultrasonic.millimiters();
  int16_t error = chain.output.regression.value - distance;

  Serial.print(" (");
  Serial.print(error);
  Serial.print(" mm off)");

  Serial.println();
  delay(1000);
}
```

As you can see, the regression chain API still consists of a single method: chain(inputs). It runs all the chained steps behind the scenes and stores the result in output.regression.value, which is a continuous number in this case. Figure 3-8 reports the logs from Listing 3-11, with the error ranging from –4 to 9 mm.

CHAPTER 3 TABULAR DATA REGRESSION

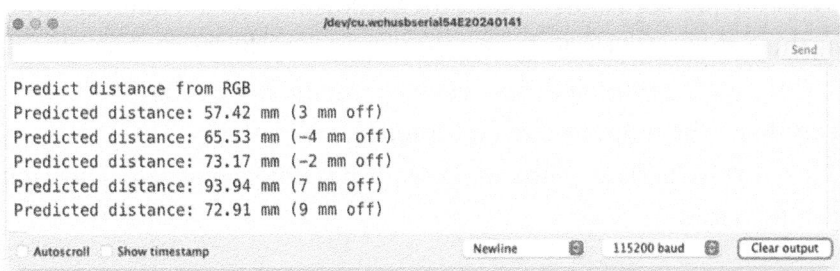

Figure 3-8. Predictions output on Serial Monitor

Now try it yourself. Start with the board near the bright, flat surface you used to capture training data and see if the predicted distance matches the measured one.

Caution   It is important that the surface and illumination are similar to when you collected the training data; otherwise, results may be off by a large amount or be completely wrong!

Summary

This chapter introduced a handful of new concepts. Tabular regression shares most of its development cycle with classification. This is why we only focused on the aspects where it differs: visualization and models.

Scatter plots were used to visualize regression data. To model regression data, there are a few new algorithms specific to this task, but there are also variations of those we already adopted for classification (decision tree and random forest).

127

After our job in Python is done, the deployment on the microcontroller using the Arduino framework stays the same as in Chapter 2. A single call to chain(inputs) runs all the necessary computation behind the scenes and produces the output we're interested in.

Now that we're done with tabular data, it is time to move to a new data type: time series.

CHAPTER 4

Time Series Classification Using Edge Impulse

Time series data refers to data where time is as important as the measured values. Rearranging the order of the samples would completely overturn the underlying data patterns. Time series classification is the task of receiving a *batch* (windows) of inputs and deciding to which class the whole batch belongs.

You can still optionally apply some transformations at the row level (normalization, binning), but feature engineering in the context of time series is focused on extracting describing statistics from each window of data.

This chapter introduces time series classification using the no-code tool Edge Impulse [1]. The project that we're going to build is a *continuous motion* classifier: it detects which gestures (among a few available) we're performing based on accelerometer data (see Figure 4-1). It serves as an archetype for many similar use cases that need to classify a constant and repetitive stream of data.

- Hourly temperature over days
- EMG (electromyography) signals from muscles
- Motor speed in an industrial setting

CHAPTER 4   TIME SERIES CLASSIFICATION USING EDGE IMPULSE

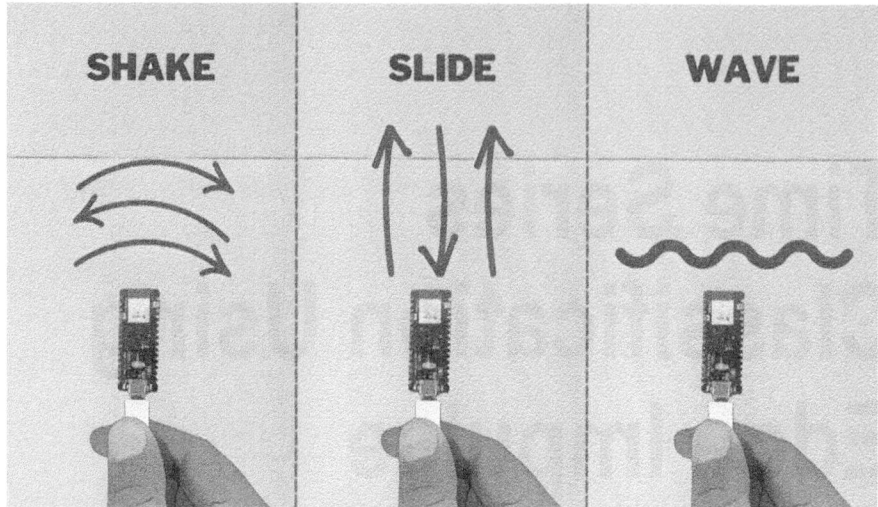

Figure 4-1. Examples of continuous gestures

The following list highlights the main steps.

1. Capture data using our microcontroller.
2. Load and inspect data on our PC using Python.
3. Create a new project on Edge Impulse.
4. Upload CSV data and specify its layout.
5. Split between train and test datasets.
6. Perform feature extraction.
7. Define and train a neural network architecture.
8. Assess the model performance on the test set.
9. Export the model into Arduino library format.
10. Deploy the Edge Impulse model back to our microcontroller.

Using a no-code tool allows you to quickly and easily iterate on various model configurations without worrying too much about the technical details. This chapter focuses on introducing time series-related concepts (windows, overlap, time-domain, and frequency-domain features) and getting the job done. Chapter 5 digs deeper into the implementation details and optimizations for a lighter and faster deployment.

Required Hardware

You need accelerometer and gyroscope data to work with. Several boards have an inertial measurement unit equipped, or you can connect an external one to your current board. Popular modules include MPU92650, MPU6886, and LSM6DS3. This book's code examples use the **Arduino Nano BLE Sense** with built-in LSM9DS1.

An *inertial measurement unit* (IMU) is a sensor module used to measure motion, orientation, and velocity. It combines data from multiple sensors—accelerometer, gyroscope, and magnetometer—to comprehensively understand movement and positioning in 3D space.

An IMU detects changes in motion and orientation using the following sensors.

- **Accelerometer**: Measures linear acceleration (e.g., movement along the x, y, and z axes). It helps determine tilt and movement speed but can't distinguish between gravitational acceleration and actual movement.

- **Gyroscope**: Measures angular velocity (i.e., rotational speed around each axis). It helps track orientation changes but is prone to drift over time.

- **Magnetometer**: Measures the earth's magnetic field to determine absolute heading (like a compass). It helps correct gyroscope drift but can be affected by local magnetic interference.

By fusing these sensors' readings through dedicated algorithms, you can estimate an object's attitude and heading state in space. IMU sensors usually come into three distinct packages.

- 3-axis: only includes the accelerometer
- 6-axis: includes accelerometer and gyroscope
- 9-axis: includes accelerometer, gyroscope, and magnetometer

The magnetometer is often overlooked, but it should still be considered if you can afford a 9-axis module since it can help mitigate the gyroscope drift error. It can also be a useful sensor when paired with a magnet (see the project in Chapter 5).

Required Software

You need the software detailed in Chapter 2 on your local computer, so please refer to that chapter if you're not following the book linearly.

Other than that, you need to create a free account on the Edge Impulse website. Head to https://studio.edgeimpulse.com/signup and complete the required steps. Figure 4-2 depicts the development workflow of the Edge Impulse platform.

CHAPTER 4   TIME SERIES CLASSIFICATION USING EDGE IMPULSE

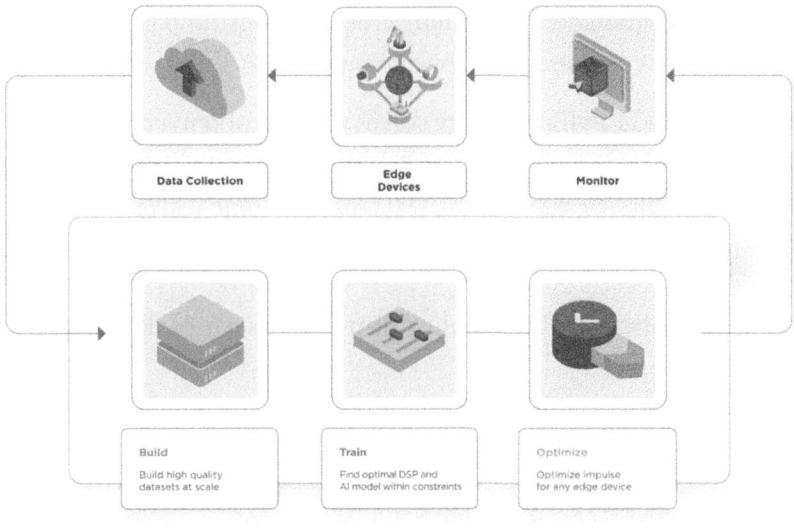

Figure 4-2. Edge Impulse development workflow

Capture Data

As showcased in Chapter 2, you have different ways to collect data from your board and move that to your PC.

- Copy-paste from Serial Monitor
- Using Python to read from Serial
- Store data on SD card

This chapter leverages method 2 and uses Python to format our time series data nicely. Copy/pasting from serial, though more immediate, suffers a couple of drawbacks.

- We're going to collect a lot of data (~100 samples/ second), and there's the risk that the Serial Monitor overflows, thus discarding old samples.

CHAPTER 4 TIME SERIES CLASSIFICATION USING EDGE IMPULSE

- Since our board doesn't have a real-time clock built-in, we would need to manually create a timestamp (we only have *relative* time by calling `millis()`).

The `tinyml4all` Python library solves these problems nicely.

Let's start with the sketch for the Arduino Nano BLE Sense. If you're using a different board, replace the IMU reading logic with the proper code (see Listing 4-1).

Listing 4-1. Collect IMU (acceleration + gyroscope) data sketch

```
/**
 * Listing 4-1: Read accelerometer + gyroscope data
 *
 * Required hardware: Arduino Nano BLE Sense.
 */
#include <Arduino_LSM9DS1.h>
#include <tinyml4all.h>

using tinyml4all::printCSV;

tinyml4all::LSM9DS1 imu;

void setup() {
  Serial.begin(115200);
  while (!Serial);
  Serial.println("Collect acc+gyro data as CSV");

  // init IMU sensor (will throw an error on failure)
  imu.begin();
}

void loop() {
  // read accelerometer and gyroscope
  imu.readAcceleration();
```

CHAPTER 4 TIME SERIES CLASSIFICATION USING EDGE IMPULSE

```
  imu.readGyroscope();

  printCSV(millis(), imu.ax, imu.ay, imu.az, imu.gx, imu.gy,
  imu.gz);

  // no manual delay, default sample rate is ~100 Hz
}
```

After you flash the sketch to your board and confirm it works by opening the Serial Monitor (see Figure 4-3), you need to run the Python script that reads the data. Listing 4-2 is similar to Listing 2-3 from Chapter 2, reported here for convenience. The only modification is in the names of columns.

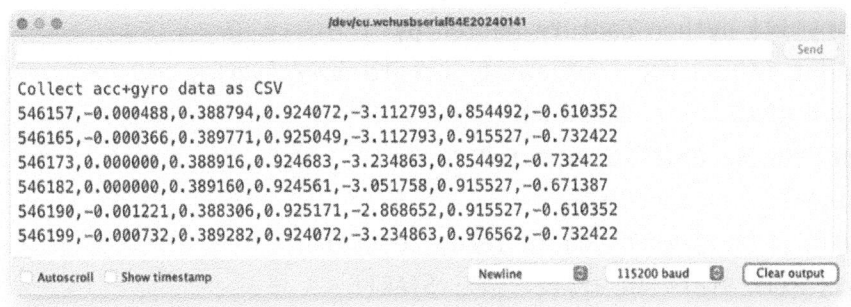

Figure 4-3. *Serial output of IMU data collection*

Listing 4-2. Capture Data from Serial in Python

```
from tinyml4all.time import capture_serial

while True:
    gesture = input("Which gesture is this? ")
    duration = input("How many seconds to capture? ")

    if not gesture or not duration:
        break
```

135

```
capture_serial(
    # * is a wildcard character
    port="/dev/cu.usb*",
    # match with Arduino sketch
    baudrate=115200,
    duration=f"{duration} seconds",
    save_to=f"motion/{gesture}.csv",
    # must match the order in the Arduino sketch!
    headings="millis, ax, ay, az, gx, gy, gz"
)
```

To run the code, open a terminal inside the folder where the script is located, activate the Python virtual environment and type.

```
(tinyml)$ python capture_motion.py
Folder motion does not exist: create now? [y|n] y
Which gesture is this? wave
How many seconds to capture? 30
Serial port connected
Press [Enter] when you're ready to start:
Task will start in 3...2...1...START!
100%|█████████████████████| 30/30 [00:30<00:00,  1.00s/it]
Collected 3689 lines of data
```

When the countdown finishes, perform your desired gesture in a continuous pattern. Good examples of continuous gestures are waving the board, moving in circles, shaking, and sliding back and forth. Repeat the collection process for every gesture you want to classify, changing the file name accordingly each time; 20–30 seconds for each gesture is a good starting point.

CHAPTER 4 TIME SERIES CLASSIFICATION USING EDGE IMPULSE

Tip The first time, start with a limited number of gestures (3–4) so that you're more likely to achieve good results without too much tuning. After you succeed, you can add more gestures later if required.

Caution Remember to record an *idle* class where you don't perform any movement; otherwise, the classifier tries to find the best match among the known gestures.

When you're done recording the gestures, you should have a folder with the different files inside your project.

```
|- your-project-root
 |- capture_motion.py
 |- motion
   |- idle.csv
   |- shake.csv
   |- slide.csv
   |- wave.csv
```

Load And Inspect the Data

Loading time series data is pretty much the same as loading tabular data, as you saw in Chapter 2 and Chapter 3. The main difference is that we're not leveraging the `Table` class. Instead, let's use the `TimeSeries` class (see Listing 4-3). Apart from that, the rest stays identical.

CHAPTER 4   TIME SERIES CLASSIFICATION USING EDGE IMPULSE

Listing 4-3. Load All CSV Files from a Folder

```
from tinyml4all.time.continuous.classification import
TimeSeries

ts = TimeSeries.read_csv_folder("motion")
ts.label_from_sources(padding="1s")
print(ts.head())
```

	timestamp	ax	ay	...	gy	gz
0	12:04:36.782244	-0,018	-0,021	...	0,977	-0,427
1	12:04:36.790244	-0,019	-0,021	...	0,977	-0,488
2	12:04:36.799244	-0,02	-0,021	...	0,732	-0,488
3	12:04:36.807244	-0,02	-0,021	...	0,671	-0,427
4	12:04:36.815244	-0,02	-0,024	...	0,916	-0,61

When loading multiple files, it is mandatory that their timestamps **do not overlap!**

Plot Time Series Data

In the context of time series data, scatter plots have no meaning. Since time is a foundational dimension, the most intuitive plot is a line with time on the *x* axis and the other measurements on the *y* axis. Listing 4-4 differentiates each measurement with a different line color and different classes by the background color, as depicted in Figure 4-4.

Listing 4-4. Draw Line Plot

```
# all the arguments are optional
ts.line(
    title="Continuous gestures",
    normalize=True,

    line_palette="magma",
    bg_palette="viridis"
)
```

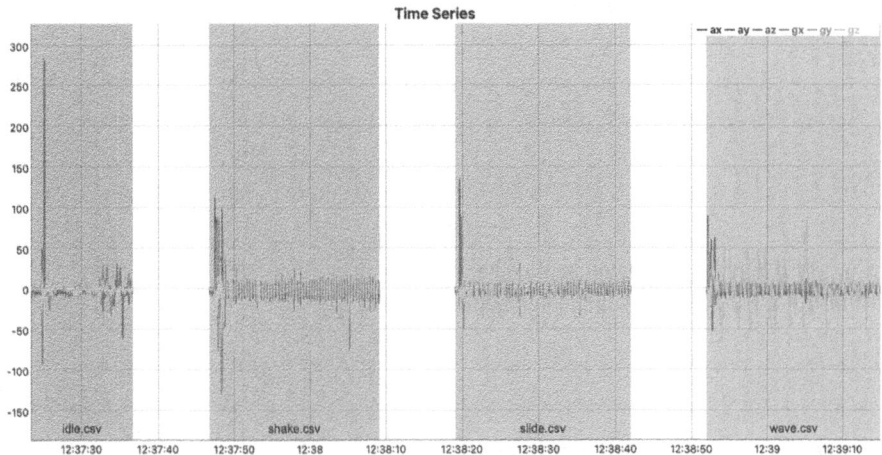

Figure 4-4. *Time series line plot*

You can interact with the plot by dragging the mouse to zoom over a specific area. Double-click to reset the zoom.

Feature Engineering

Time series feature engineering is focused on *windows* of data. A window is simply a group of readings related by their timestamps.

A window is a FIFO (first in, first out) data structure: when a new sample arrives, it is *queued* at the end of the window. When the window reaches its maximum length, the eldest sample is discarded to make room for the new one (see Figure 4-5).

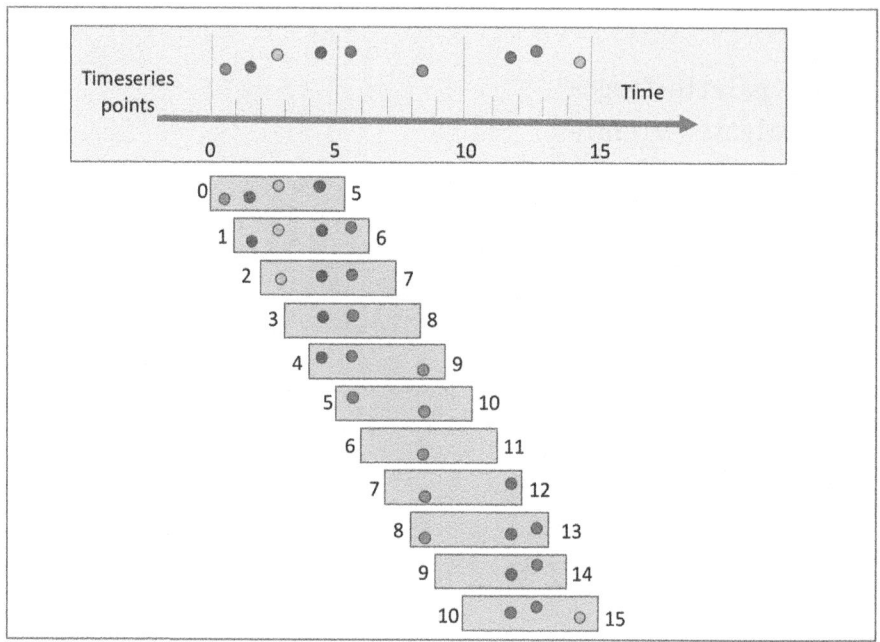

Figure 4-5. *Rolling window logic*

Figure 4-5 depicts the most general case, where each window has the same *duration*, but the number of samples it contains may vary (due to a signal with non-constant frequency). In our case, though, the machine learning model expects a fixed number of inputs, so there are two choices.

- Ignore the windows that do not group the correct number of samples.

- Consider windows of the same *length* instead of the same duration.

Option 1 suffers from several problems (e.g., if data is produced at irregular intervals, we risk never—or hardly ever—getting a window with the correct number of samples). Option 2 is easy to implement and guarantees the generation of predictions as new samples arrive (the drawback is that it assumes a constant sampling frequency—which may not always be the case). For ease of implementation, from now on, let's assume that all windows of data share the same number of samples. This number is the *length* of the window.

Let's make a practical example. Suppose you have an array of ten values.

2, 3, 6, 4, 3, 4, 7, 8, 5, 4

We can chunk this data into windows of five items, with a shift (how much the window slides to the right) of 2. This results in windows like the following lines display.

```
window 1 (index 1 to 5): 2, 3, 6, 4, 3
window 2 (index 3 to 7): 6, 4, 3, 4, 7
window 3 (index 5 to 9): 3, 4, 7, 8, 5
```

Given a window of data, we can generate features in two domains.

- **Time** analyzes the data "as is" as a sequence of numbers over which you can compute statistical moments (min, max, average, standard deviation, etc.) and shape descriptors (number of peaks, count of values above or below the mean or zero, etc.)

- **Frequency** applies a transformation (Fourier transform [2]) that maps the original sequence of numbers into a new domain. In this domain, the numbers represent how much a given frequency is relevant to the original signal.

There's also a third option: don't generate features at all. Or, better, let a neural network figure out the features on its own. You see this in action in Chapter 8.

Each domain has advantages and disadvantages.

Time-Domain Features

If you treat each measurement from the window of data as a numeric series, you can borrow a lot of metrics from classic statistics to describe the series. The overall goal of all these metrics is to describe the *shape* of the series (see Figure 4-6). Metrics like mean, standard deviation, skew, and kurtosis are often sufficient to capture the intrinsic characteristics of a time series and help differentiate among different classes.

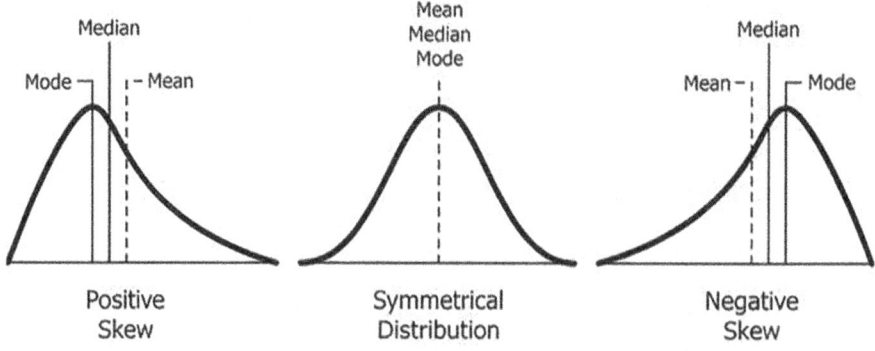

Figure 4-6. *Example of skewed curves*

One of the advantages of time-domain features is that they make very few assumptions about the data patterns. You can get meaningful metrics over any series, no matter how noisy or irregular it is; it can even handle missing data pretty well. The second advantage is that these metrics are easy and fast to calculate: there's no complex math involved, and any embedded CPU can complete this task in the order of microseconds.

On the other hand, this simplicity can sometimes become a limitation, and the descriptive power of these metrics could not be sufficient to capture the underlying data patterns, especially in the case of repetitive, periodic time series.

Frequency Domain Features

Frequency domain features don't look at the raw numbers in the series. They apply a transformation to project the time values into the domain frequency. This operation is performed by the Fourier transform. The theory and math behind it can be a little daunting to get right and is out of the scope of this book. What matters to us in tiny machine learning is that this operation requires much more computations than the classic statistic features. The features it extracts are meant to describe the main frequencies that compose a time series instead of its shape (see Figure 4-7).

CHAPTER 4 TIME SERIES CLASSIFICATION USING EDGE IMPULSE

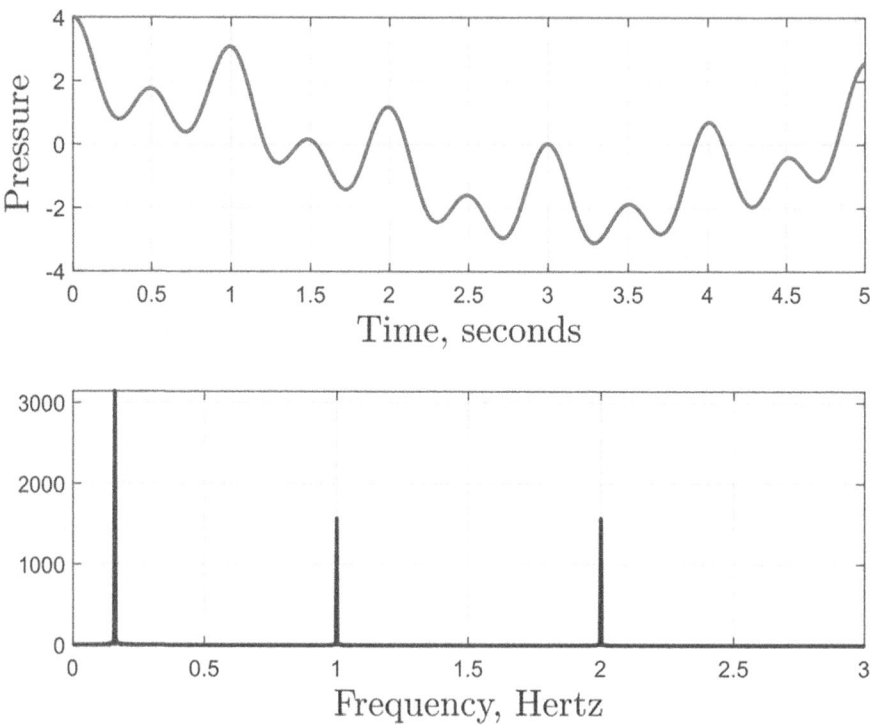

Figure 4-7. Frequency content of time series

So, why would you prefer the frequency domain over (or in conjunction with) the time domain?

The answer is that frequency domain features exhibit superior descriptive power in the context of repetitive and periodic patterns. They're usually much more *dense* (produce a higher number of coefficients) and can discriminate with confidence between different classes. Fourier transform is a great choice if you're modeling continuous motion, like in this chapter.

CHAPTER 4 TIME SERIES CLASSIFICATION USING EDGE IMPULSE

Edge Impulse for Continuous Motion

This section is a step-by-step walkthrough of the Edge Impulse platform for continuous motion classification. Edge Impulse can also handle audio data and images, as explained in Chapter 6 and Chapter 7. Getting familiar with this tool speeds up your workflow across many different projects, so take some time to explore it beyond this chapter's strict requirements.

Edge Impulse is a *low-code* tool: it allows you to generate a model without writing any code through their GUI. You only need to write code to integrate the generated model inside your Arduino sketch. It is a great tool for beginners who don't want to dig too much into the complex details of feature extraction and classification. It also provides built-in quick tools for data labeling, so you don't have to use third-party software for this task.

Head to `https://studio.edgeimpulse.com/studio/profile/projects` and create a new project. Call it `continuous motion`.

> **Note** You can give your project any other name you prefer. (Keep it under 20 characters, though, since long names may cause troubles later with the Arduino IDE). Remember that you need to adapt some code later to reflect the change!

Edge Impulse Workflow

When working with the Edge Impulse platform, your workflow is very linear. The side menu on the project home page highlights all the available steps in (almost) the exact order you must follow (see Figure 4-8).

- **Dashboard**: An overview of the project, with quick links to later steps, tutorials, and project management actions (e.g., delete project).

CHAPTER 4   TIME SERIES CLASSIFICATION USING EDGE IMPULSE

- **Devices**: You can connect one of the supported devices using Serial (via EI CLI), WebUSB, or Wi-Fi (e.g., your smartphone) and capture data directly from this page without first creating a CSV file, as we've done so far. I don't usually use this feature since I prefer to have the files I'll be working on in the format I prefer, but this can come in handy for quick tests.

- **Data acquisition**: Here, you can handle your datasets. It keeps track of the files you upload, handles the train/test split, and gives a preview of the data (it's really handy for images!). You can perform many actions (delete, label, move samples) with the context menu.

- **Impulse design**: This is the *core* of the entire platform. An *impulse* is the equivalent of the chain seen in Chapter 2 and Chapter 3. It groups the feature engineering process and a classification (or regression) model. Its job is to take your input data and perform the computations required to obtain the expected output. It is articulated in nested steps.

- **Create impulse**: A configuration screen where you specify the type of task and features you want to use (classification vs. regression, spectral features vs. raw data.)

- **Feature extraction**: The name of this menu item changes according to what you selected in the previous screen. For time series classification, this item will likely be *spectral features*. No matter the specific item, though, this page allows configuring the specific parameters of the feature extractor step.

- **Classifier**: Here, you define the neural network topology (either from scratch or using a reference architecture among the supported ones) and run the training. When the process is completed, you get a nice confusion matrix to inspect the results.

- **Live classification**: If you connected a supported device earlier, you can now run predictions over the live data from the device without the need to flash any firmware. In this case, the predictions are made in the cloud, so you can preview the model's accuracy on real-world data.

- **Model testing**: Recall that the *data acquisition* step handles the train/test split. On this page, you can run the model on the test dataset and see how it performs.

- **Deployment**: After you're done with the configuration, tuning, and testing online, it's time to export the trained model to C++ so that it can be integrated into your project. This page gives many export options for the many supported boards and desktop operating systems, too.

- **Versioning**: Frequently, you experiment with different datasets, different models, and different configurations. You may want to keep track of all these experiments to avoid losing important settings that you may want to recall later. Versioning gives the possibility to create a *snapshot* of the current state of the project in time. You can then make the edits you want with the peace of mind that you will later be able to restore the previous state as if nothing happened later.

CHAPTER 4   TIME SERIES CLASSIFICATION USING EDGE IMPULSE

- **Experiments**: It is a more advanced feature that allows you to configure and train multiple impulses over the same data and compare their results to select the best. By default, it only lists the results of the current (only) impulse, but as you create new ones, they are listed in a nice table for quick comparison.

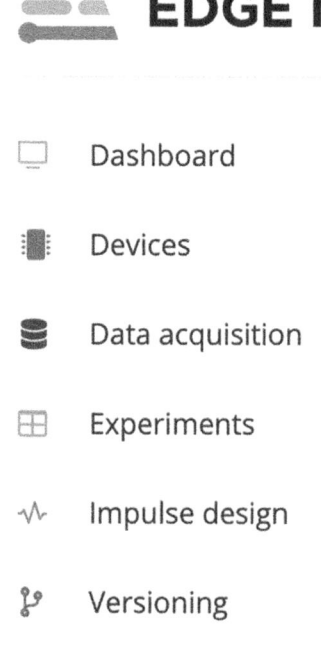

Figure 4-8. Edge Impulse project menu

The projects in this book don't use every feature of the platform. The following describes the steps of our workflow.

1. Configure data format (this is only necessary in this chapter since data in CSV files can be arranged in different layouts. Images and audio files are self-explanatory).

2. Upload data and split into train/test.
3. Define the impulse structure and parameters.
4. Perform feature extraction.
5. Train a model.
6. Deploy as an Arduino library.

CSV Layout Configuration Wizard

Edge Impulse is often used for audio and image classification. Audio and image files are self-explanatory in that each file clearly means what it represents (and how this information is encoded). CSV data, on the other hand, can be arranged in many different layouts, and each row may contain a wide range of data types. For this reason, you have to instruct Edge Impulse about how to interpret the files you're going to upload before it can ingest them properly.

You perform this operation using the CSV Wizard, a guided configuration tool that incrementally asks you to define the layout of your data.

Tip Refer to the online materials for a video walkthrough of the CSV Wizard configuration.

Navigate to **Data acquisition ➤ CSV Wizard** tab and upload one of the files you captured earlier (see Figure 4-9).

CHAPTER 4   TIME SERIES CLASSIFICATION USING EDGE IMPULSE

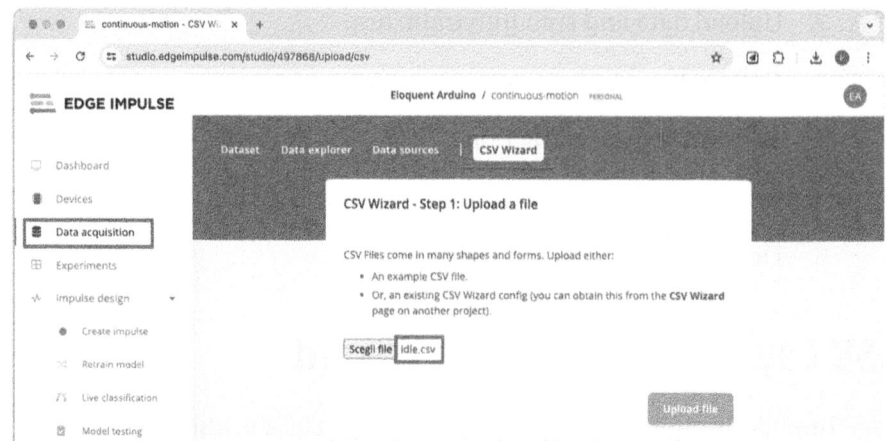

Figure 4-9. Edge Impulse CSV Wizard intro

Confirm the data looks correct and move to step 3 (see Figure 4-10). Here, you must select the following options.

- **Is this time series data?** Yes, this is time series data (either raw sensor data or processed features).

- **How is your time series data formatted?** Each row contains a reading, and sensor values are columns.

- **Do you have a timestamp or time elapsed column?** Yes, it's timestamp.

- **What type of data is your timestamp column?** Full timestamp.

- **Override timestamp difference?** *<your sampling frequency>*. If it warns you that the difference is not consistent, choose the proper one. For example, in our project, the sampling rate is 120 Hz, so the time interval is 8.3 ms. You can round to 9 or 8; it makes no difference.

CHAPTER 4   TIME SERIES CLASSIFICATION USING EDGE IMPULSE

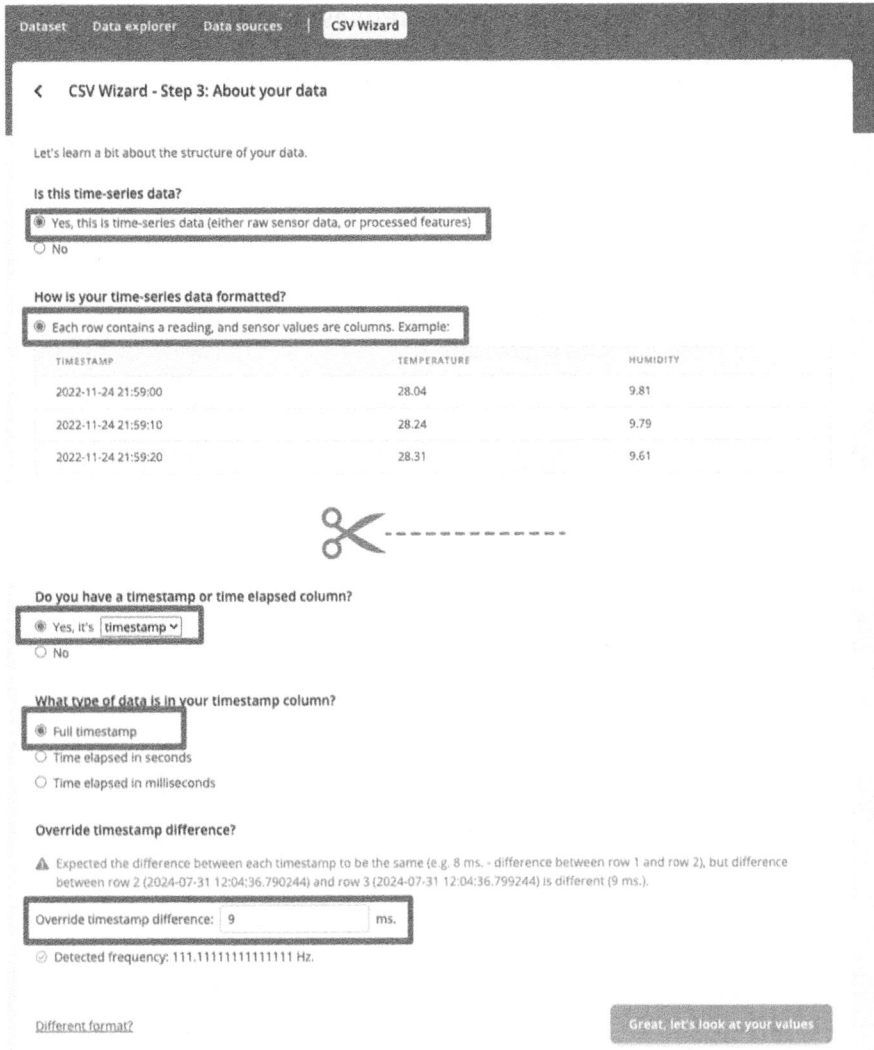

Figure 4-10. *Edge Impulse CSV Wizard*

The next two steps are short and easy (see Figure 4-11).

- **Do you have a column that contains the label (the value you want to predict)?** No.

- **Which columns contain your values?** Tick all.

CHAPTER 4   TIME SERIES CLASSIFICATION USING EDGE IMPULSE

- **How long do you want your samples to be?** Unlimited.

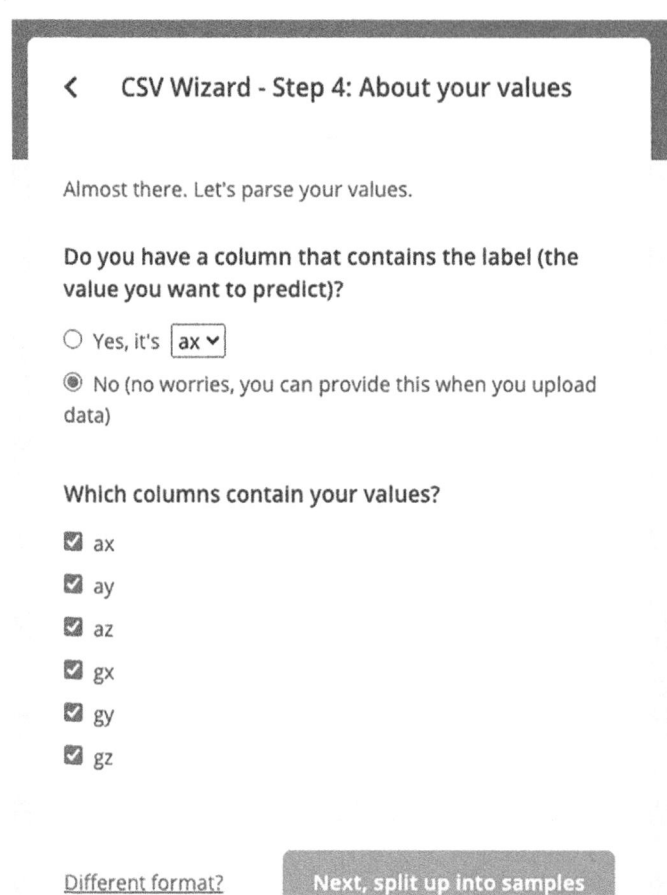

Figure 4-11. Edge Impulse CSV Wizard configuration (continued)

Now that the data layout has been defined, you can import as many files as you need without hassle.

CHAPTER 4   TIME SERIES CLASSIFICATION USING EDGE IMPULSE

Upload Files

Head back to **Data acquisition** from the menu and click the **Add data** button. Choose **Upload data** and select the files you captured in the previous step. Leave the default options checked (see Figure 4-12).

- Select individual files
- Training
- Infer from the file name

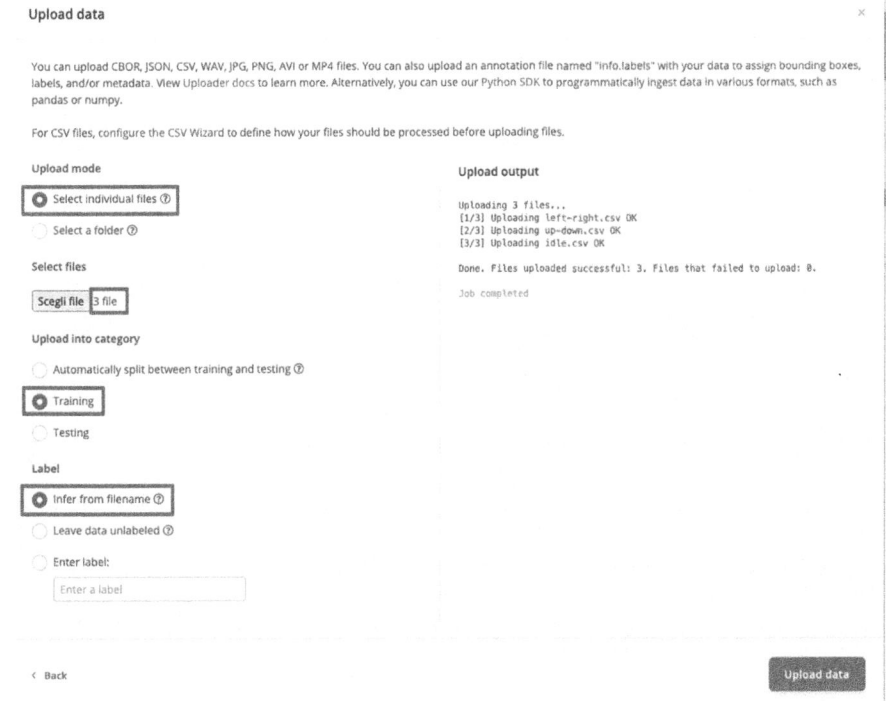

Figure 4-12. Edge Impulse data upload form

153

We uploaded all files as training data because Edge Impulse performs train/test splitting on CSV files by applying an automatic split that doesn't preserve the correct time order of data and can't be configured. Manual splitting gives you greater control over the process, such as deciding the split percentage.

Train/Test Split

You should have a few entries in your dataset panel, one for each file. Go through each row and click the three dots on the right. This opens a context menu with a list of actions. Click **Split sample** to open the split GUI. Delete all the segments that have been created automatically and only define two segments: one for training and one for testing. The first (for training) should be larger so that the model has more data to learn from. Figure 4-13 shows an example.

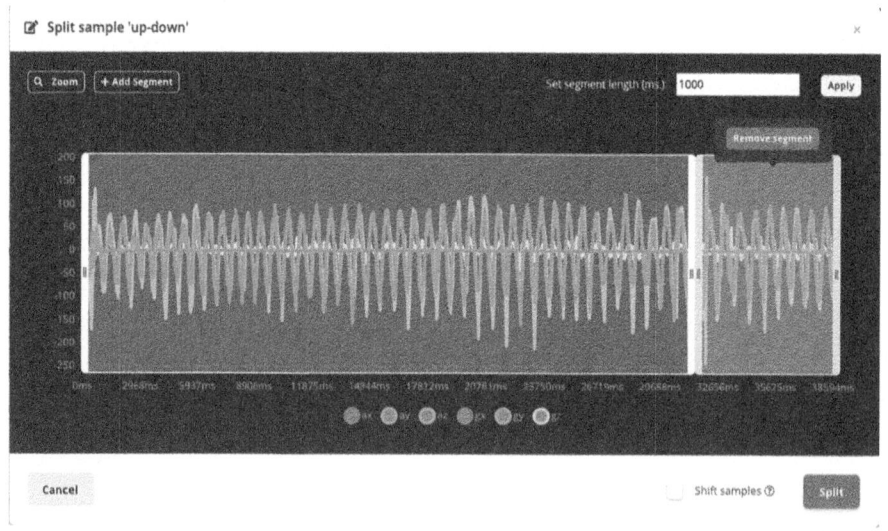

Figure 4-13. *Edge Impulse train/test split*

CHAPTER 4   TIME SERIES CLASSIFICATION USING EDGE IMPULSE

Returning to the Data Acquisition page, you should see one new row in the list of samples. The naming follows the scheme <label_of_class>.s1 and <label_of_class>.s2 (e.g., idle.s1 and idle.s2). s1 is the sample to use for training; s2 is for testing. Click the three dots on the right of the s2 sample and select **Move to test set**. Then, repeat the same procedure for all your uploaded samples (see Figure 4-14).

Figure 4-14. Edge Impulse dataset statistics

Impulse Design

Here's where the actual classification tasks start. Each type of data has its own feature engineering block, which you can configure visually using the GUI (see Figure 4-15).

155

CHAPTER 4   TIME SERIES CLASSIFICATION USING EDGE IMPULSE

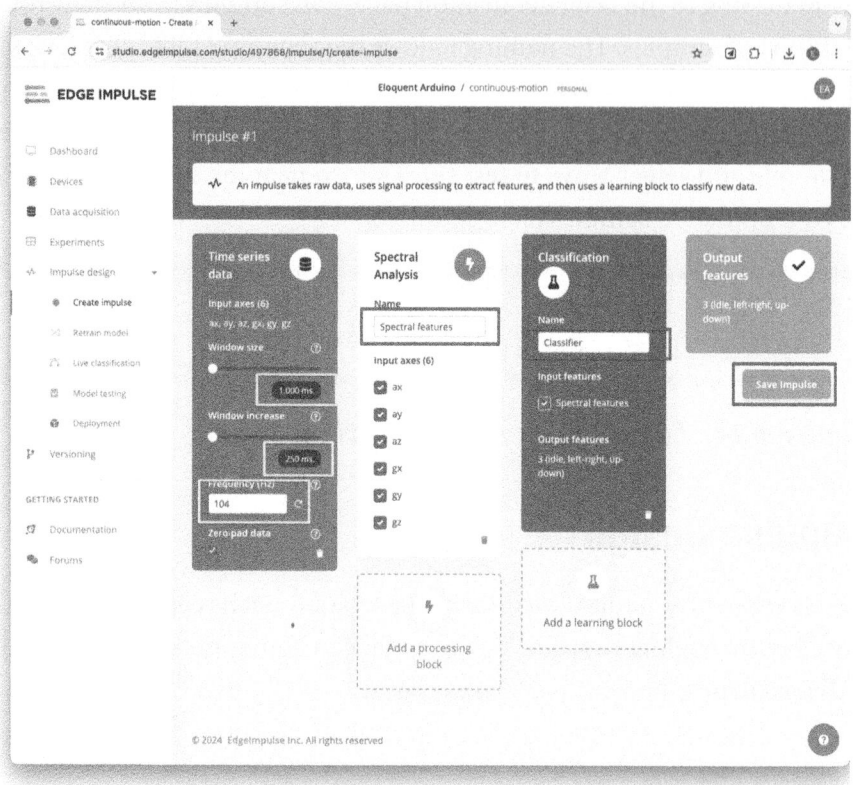

Figure 4-15. Impulse design

Time Series Data Block

In our case of time series classification, two parameters need to be configured.

- **Window size**: The duration of each window of data. It must be configured in milliseconds because the Edge Impulse platform assumes your data has a fixed sampling rate (inferred from the uploaded data).

CHAPTER 4   TIME SERIES CLASSIFICATION USING EDGE IMPULSE

- **Window increase**: This was called *shift* earlier in this chapter and indicates how much (again in milliseconds) the window should slide to the right when it is full.

In continuous motion, you should size the window duration to capture at least a few repetitions of the gestures you want to recognize. For the sake of this project, a value between one and two seconds works fine.

Spectral Analysis Block

As a preprocessing block, select **Spectral Analysis** and tick all the input axes. Under the hood, this block extracts a mixture of time domain (like root mean square value, skew, and kurtosis) and frequency domain features (spectral power at different frequencies).

After you save the impulse by clicking the respective button, a **Spectral features** entry appears on the left menu, where you can configure many parameters for this block. To be tuned properly, they required a strong understanding of how FFT and other frequency domain algorithms work—which is out of the scope of this book—so for the moment, it is fine to continue with the defaults (see Figure 4-16). If you later feel that the model is not performing at its best and want to dig more into this page, the most intuitive parameter you can tweak is the *FFT length;* increasing this value considers more points for the FFT generation and *could* potentially improve the overall accuracy.

CHAPTER 4   TIME SERIES CLASSIFICATION USING EDGE IMPULSE

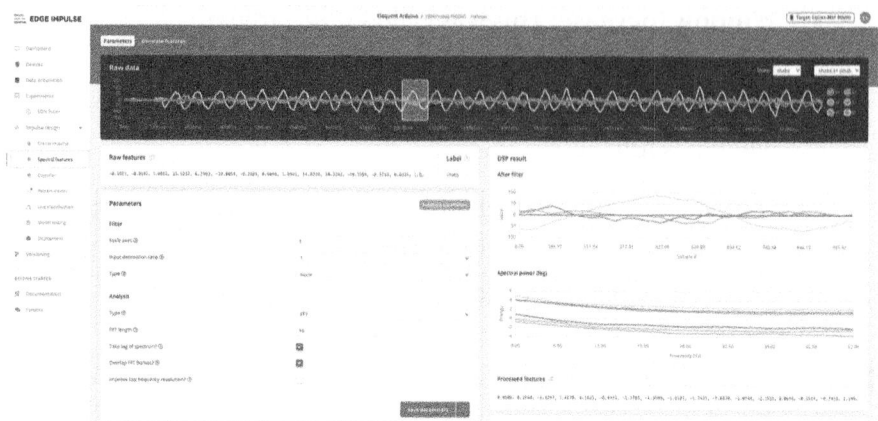

Figure 4-16. Edge Impulse feature extraction configuration

Next, click **Save parameters** to go to the Generate Features page. Click the **Generate features** button and wait for the process to finish (this step is mandatory; otherwise, you won't be able to train the model later). When it has been processed, a scatter plot appears, displaying the clustered data (similar to the scatter plot drawn in Chapter 2 for tabular data). Figure 4-16 shows the distinct colored clusters for each class.

Learning Block

The last block to configure is the *learning* block. It is pretty limited right now and it even guides us by showing the recommended block that is *classification*, so select that and hit the **Save Impulse** button.

A *classifier* entry appears in the menu on the left, under **Spectral features**. Click it to open the model training configuration page. The default topology proposed is a neural network of two *dense* layers with few neurons each. On my dataset, leaving all the options to their default value achieved near 100% classification accuracy, indicating that the extracted features are so good that even a simple model without any proper customization can correctly classify them (see Figure 4-17).

CHAPTER 4 TIME SERIES CLASSIFICATION USING EDGE IMPULSE

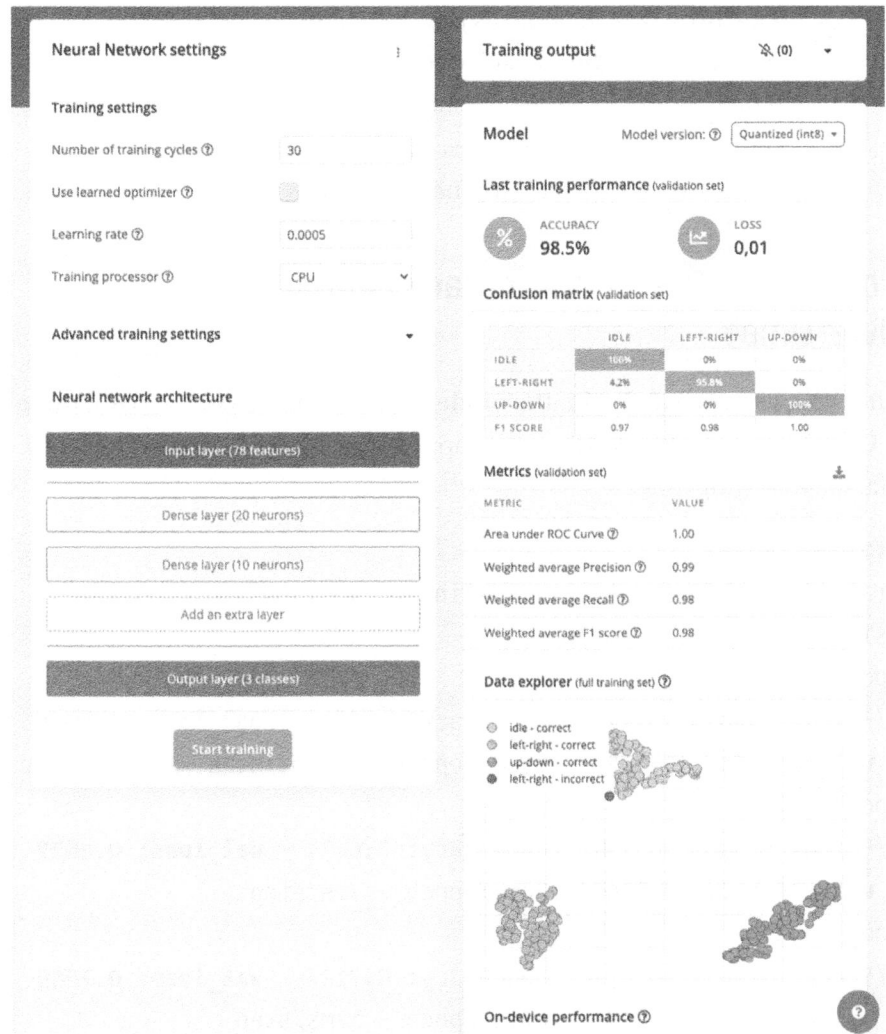

Figure 4-17. Edge Impulse classification

You can use a couple of important parameters a lot while working on this page, so here's what they do.

- **Number of training cycles**: It addresses how long the model should learn. A sensible default value is 50. You can try to increase this value if the accuracy is low, but it looks like it is increasing (see later for how to check this).

- **Learning rate**: It controls how much the model should update its weights at each epoch. My go-to value is 0.005: lower values may result in higher accuracy but require more training cycles. Notice that it differs from the 0.0005 set by default (one less zero)!

How Do We Detect If Accuracy Is Improving Over Time?

After you click **Save & Train**, the model starts learning. On the right, there's a **Training output** block where logs appear during the process. They look like the following lines.

```
Training model...
Training on 157 inputs, validating on 40 inputs
Using batch size: 32
Epoch 1/30
5/5 - 1s - loss: 1.2153 - accuracy: 0.6624 - val_loss: 0.8915
- val_accuracy: 0.6750 - 831ms/epoch - 166ms/step
Epoch 2/30
5/5 - 0s - loss: 0.7945 - accuracy: 0.6433 - val_loss: 0.4627
- val_accuracy: 0.6750 - 34ms/epoch - 7ms/step
Epoch 3/30
5/5 - 0s - loss: 0.4370 - accuracy: 0.7134 - val_loss: 0.2049
- val_accuracy: 0.9500 - 62ms/epoch - 12ms/step
Epoch 4/30
5/5 - 0s - loss: 0.2389 - accuracy: 0.9045 - val_loss: 0.1047
- val_accuracy: 1.0000 - 29ms/epoch - 6ms/step
Epoch 5/30
5/5 - 0s - loss: 0.1518 - accuracy: 0.9554 - val_loss: 0.0726
- val_accuracy: 1.0000 - 61ms/epoch - 12ms/step
```

As you can see, each line reports the accuracy and loss on the validation split at each epoch. You already know what the accuracy is, but what about the loss? It is a function that quantifies how well or poorly a neural network is performing by measuring the difference between the model's predictions and the actual target values. If the loss decreases, the model learns better representations of the data. During the training phase, the objective of the network is to minimize this value rather than increase accuracy. An high Model accuracy is a direct result of a model with a low loss value—because it has learned a good representation of the data, it can produce good predictions that match the ground truth.

In the example, the validation loss is decreasing at each step: that's a sign that the model is learning. (It is expected since we're beginning the process.) If you look at the logs from the last few epochs, you can see that the loss stays almost the same.

```
Epoch 57/60
5/5 - 0s - loss: 0.0092 - accuracy: 1.0000 - val_loss: 0.0022
- val_accuracy: 1.0000 - 61ms/epoch - 12ms/step
Epoch 58/60
5/5 - 0s - loss: 0.0090 - accuracy: 1.0000 - val_loss: 0.0022
- val_accuracy: 1.0000 - 29ms/epoch - 6ms/step
Epoch 59/60
5/5 - 0s - loss: 0.0088 - accuracy: 1.0000 - val_loss: 0.0021
- val_accuracy: 1.0000 - 61ms/epoch - 12ms/step
Epoch 60/60
5/5 - 0s - loss: 0.0087 - accuracy: 1.0000 - val_loss: 0.0021
- val_accuracy: 1.0000 - 28ms/epoch - 6ms/step
```

That is a sign that we're running the training for too many cycles and wasting time because the model reached a *plateau*. Even more, it may be the case that, for certain datasets, the model starts overfitting, resulting in a degradation of its accuracy!

> **Caution** When training a model, keep an eye on the validation loss to spot signs of overtraining and overfitting!

Chapter 8 delves into the different neural network architectures, but let's briefly go over what a fully connected neural network (FCNN) is and what the *dense* layers in Edge Impulse represent.

Fully Connected Neural Networks

Artificial neural networks are data structures inspired by the structure and function of the human brain. A neural network is a complex system of interconnected nodes or *neurons* that process and transmit information. Each neuron receives one or more inputs, performs a computation on those inputs, and then sends the output to other neurons. This process allows the network to learn and represent complex patterns in data. Neurons (the atomic component) are organized into *layers,* stacked one after the other. Fully connected neural networks are characterized by each neuron at layer *n* being connected with each neuron at layer *n* + 1 (see Figure 4-18).

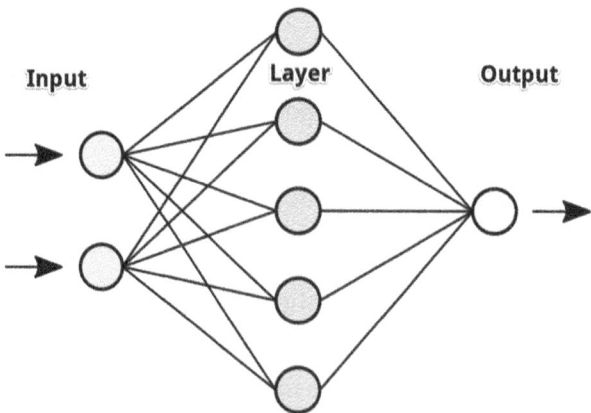

Figure 4-18. *Fully connected neural network*

The Structure of a Perceptron

A neuron, also known as a *perceptron*, is the basic building block of a fully connected neural network. It consists of four main components.

- **Inputs**: The neuron receives one or more input values from the previous layer (or the input sample in the case of the first layer).

- **Weights**: Each input is associated with a weight, which determines the importance of that input in the calculation.

- **Bias** (optional): Each neuron may introduce a fixed offset to shift its output value.

- **Activation function**: When working with linear weights, this function introduces a non-linearity, which allows us to learn more complex relationships.

The output of the neuron is calculated using the following formula.

$$output = activation\left(\sum(inputs \cdot weights) + bias\right)$$

What Are Dense layers?

A dense layer is a group of perceptrons defined by the number of neurons in it. Considering the full connection, you have to remember that the number of weights of a dense layer grows linearly with the number of its inputs and neurons. A dense layer with 20 inputs and 30 neurons stores 600 weights.

Now that you know what a dense layer is and how the number of neurons influences the size of the network feel free to experiment with different combinations of layers until you find the sweet spot between size (lower is better) and accuracy (higher is better).

CHAPTER 4   TIME SERIES CLASSIFICATION USING EDGE IMPULSE

Testing

If the confusion matrix on the training page satisfies your expectations, move to the Model Testing page (from the menu on the left) to assess the model accuracy on samples it has never seen during training or validation. Click the **Classify all** button and wait for the results. In the best scenario, all the rows are green (indicating a perfect prediction). A realistic scenario like ours probably gets results that look worse than those during training. That's totally fine and expected.

Remember that the accuracy you get on the validation data is an overestimation of the model's true capability to generalize to new data, so this page is of utmost importance in deciding whether you can be satisfied with your model. It may well happen that a 100% accuracy on the training page becomes 80% or lower on the test set. It may be a sign of overfitting, and you need to take action to counter it (e.g., tuning the Spectral Features step or the network architecture).

In my case, the accuracy on the test set was still 100%, so I moved on to deployment.

Deployment

The training procedure is complete at this point, and you can move your model from the cloud to your device. Click **Deployment** in the left menu and export the impulse as an Arduino library (see Figure 4-19).

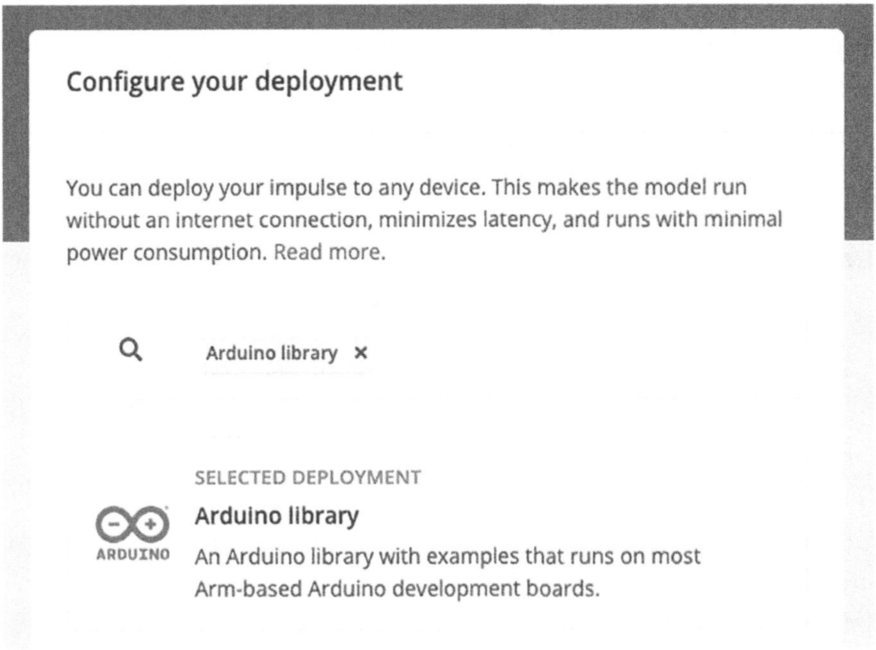

Figure 4-19. Edge Impulse export

This downloads a ZIP file. Extract the contents of the zip inside your Arduino IDE libraries folder. The library is named <name-of-project_inferencing> (continuous-motion_inferencing in my case).

Caution If you want to rename the library, remember to rename both the folder *and* the .h file inside the src folder.

Now that the Edge Impulse model is a library, it is time to modify the initial sketch that only captured data to integrate the classification part. Create a new sketch called Continuous_Motion_Classification.ino and paste the code shown in Listing 4-5.

CHAPTER 4   TIME SERIES CLASSIFICATION USING EDGE IMPULSE

> **Caution** Listing 4-5 assumes your library is named `continuous-motion_inferencing`. If your generated library has a different name, replace it with the correct one.

Listing 4-5. Run Impulse on Accelerometer Data

```
/**
 * Listing 4-5: Run Edge Impulse model on IMU data
 *
 * Required hardware: Arduino Nano BLE Sense
 */
// next line *must* match with the window increase value in the
Impulse Design page from EI Studio
#define EI_CLASSIFIER_SLICES_PER_MODEL_WINDOW (1000 / 250)

#include <continuous-motion_inferencing.h>
#include <Arduino_LSM9DS1.h>
#include <tinyml4all.h>

using tinyml4all::printCSV;

tinyml4all::LSM9DS1 imu;
tinyml4all::Impulse impulse;

void setup() {
  Serial.begin(115200);
  while (!Serial);
  Serial.println("Classify motion with Edge Impulse");

  imu.begin();

  // init EI model (will throw an error on failure)
  impulse.begin();
}
```

CHAPTER 4 TIME SERIES CLASSIFICATION USING EDGE IMPULSE

```
void loop() {
  // read accelerometer and gyroscope
  imu.readAcceleration();
  imu.readGyroscope();

  // queue data
  impulse.queue(
    imu.ax, imu.ay, imu.az,
    imu.gx, imu.gy, imu.gz
  );

  // when window of data is full, run prediction
  // otherwise return
  if (!impulse.isReady())
    return;

  Serial.print("Running prediction... ");

  // run inference
  if (!impulse.run()) {
    Serial.println(impulse.error());
    return;
  }

  // print results
  Serial.print("Predicted motion is ");
  Serial.print(impulse.label());
  Serial.print(" with confidence ");
  Serial.println(impulse.confidence());
}
```

Compile and upload the sketch, then open the Serial Monitor to see the live stream of predictions (see Figure 4-20).

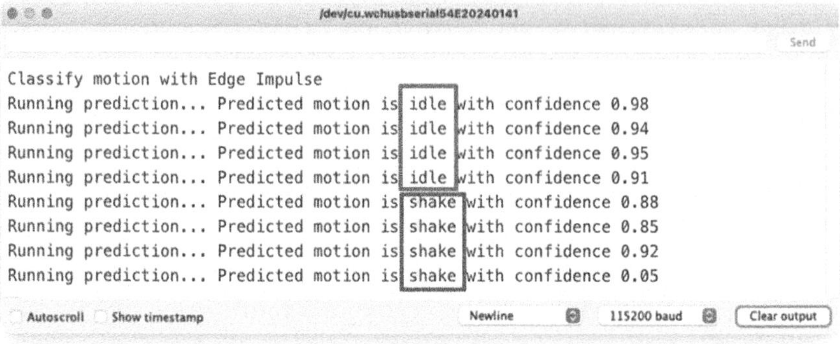
/dev/cu.wchusbserial54E20240141
Send
Classify motion with Edge Impulse
Running prediction... Predicted motion is idle with confidence 0.98
Running prediction... Predicted motion is idle with confidence 0.94
Running prediction... Predicted motion is idle with confidence 0.95
Running prediction... Predicted motion is idle with confidence 0.91
Running prediction... Predicted motion is shake with confidence 0.88
Running prediction... Predicted motion is shake with confidence 0.85
Running prediction... Predicted motion is shake with confidence 0.92
Running prediction... Predicted motion is shake with confidence 0.05
Autoscroll
Show timestamp
Newline
115200 baud
Clear output

Figure 4-20. *Inference serial output*

Caution Compilation time may be long. On my Intel MacBook Pro, it takes three to four minutes!

Congratulations! You've completed your first Edge Impulse project.

Hopefully, it was a streamlined process that you can easily replicate in future projects. If you ever feel lost, return to this chapter and use it as a reference and step-by-step guide.

Note The generated model takes ~90 milliseconds to run on my Arduino Nano BLE Sense. This value is used as a benchmark in the next chapter, which implements only time-domain features to see how they compare in terms of execution time.

Visit the online materials repository for a video demo of the live classification.

Edge Impulse Shortcomings

Edge Impulse is a great platform. I frequently use it, and I'm amazed by the quality of the models it produces. That said, there are a few shortcomings that still frustrate me (not on the platform itself, but on the Arduino IDE side).

Painfully Long Compilation Times

Edge Impulse uses TensorFlow under the hoods. This means that every time you compile your project, TensorFlow needs to be compiled, too. Compilation takes time because it has a large codebase on its own. A lot of time. This can quickly become annoying when you're in the development stage and need to make frequent changes to code. Even after the first compilation, when the Arduino IDE has created some cache artifacts to speed up the process, compiling your sketch may take about one minute.

Cache Invalidation

Let's say you train a model, deploy it to your board, and find it performs badly. Then, you return to Edge Impulse and train a new model with different configurations to improve its accuracy. You download the new ZIP library, replace the old one, and compile your sketch again.

Guess what? You're still using the old model! As counterintuitive as it is, that's what's happening. This is because the Arduino IDE builds a cache to speed up the compilation process. This cache lives until you restart the IDE. Due to the caching implementation, the IDE cannot detect that the files from your library have changed (since they all have the same names), so it continues reusing the cached version. To force the use of the new version, you must restart the IDE altogether. This means that you must wait.

- The time it takes to launch the IDE
- The time it takes to compile TensorFlow for the first time (3–4 minutes)

CHAPTER 4 TIME SERIES CLASSIFICATION USING EDGE IMPULSE

I don't know about you, but this waiting destroys my productivity.

All that said, I still think Edge Impulse models are a great option for your TinyML projects.

Summary

Time series is a common data type when working with TinyML projects. It covers every project where data changes over time (accelerometer, temperature, EMG), and being able to classify this data opens a wide range of opportunities.

This chapter presented the Edge Impulse platform, a low-code tool that empowers you to visually create advanced machine learning models ready to be deployed. Under the hood, it uses frequency domain features (Fourier transform) and TensorFlow, thus achieving top-notch accuracy. Thanks to the `tinyml4all` Arduino library, it only takes a couple of lines of code to integrate these models into your Arduino projects.

The next chapter demonstrates how it is possible to classify time series data using time-domain features and "classical" machine learning models. This produces a more lightweight classification pipeline that is powerful enough to achieve satisfactory results.

CHAPTER 5

Time Series Classification Using Python

Chapter 4 introduced the problem of time series classification. By leveraging a low-code tool like Edge Impulse, we successfully deployed a classification model without digging into the details of Fourier transform and power spectrum feature extraction. Also, the neural network used as a classifier acted like a black box to us, leaving us only the mandatory task of configuring the number and width of layers (you are even allowed to configure the Python code to define and train the model, but that is in "Expert" mode and out of the scope of this introductory book).

This chapter takes a more *code-oriented* approach and explores what makes a good time series classification pipeline using time-domain features, introducing new feature engineering operators. As a last step, we export the selected chain to plain C++ code (like in Chapters 2 and 3) to be embedded into any Arduino project.

The project of this chapter is a media control pad for our PC.

The following are common actions you expect from a media control device.

- **Raise/lower volume**: Slide your finger up (raise) or down (lower) along the imaginary vertical axis of the pad.
- **Move next/back**: Slide your finger left (next) or right (back) along the imaginary horizontal axis of the pad.
- **Play/pause**: Tap the middle of the pad.

Let's perform these gestures on a flat surface (e.g., a desk) with the help of a small magnet (see Figure 5-1).

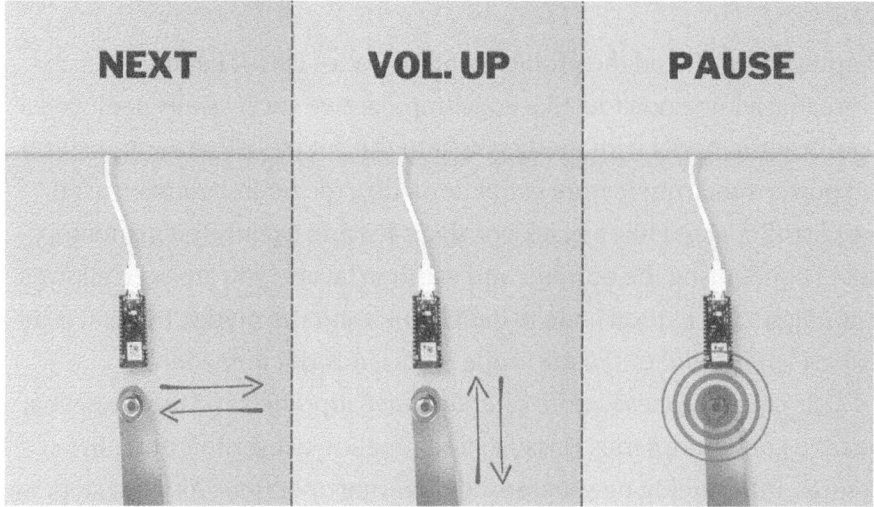

Figure 5-1. Examples of gestures

Hardware Requirements

You could leverage accelerometer and gyroscope data in this project, like in Chapter 4, but let's try to vary the nature of the data we are working with. This time, we exploit the Arduino Nano BLE Sense built-in magnetometer instead. Many of the inertia measurement unit (IMU) sensors mentioned in Chapter 4 have such a sensor built-in, so you don't need external hardware. A little magnet taped at the tip of your finger can disturb the earth's magnetic field in a measurable way; that perturbation (on the x, y, and z axes) becomes our input data.

For this project, you need the following.

- An IMU with a magnetometer (e.g., the one on the Arduino Nano BLE Sense) or an external magnetometer sensor
- A small magnet
- A flat surface to use as a pad

Tip Refer to the online materials repository for a video demo of the setup.

Capture Data

This section shares a lot with the similar one in Chapter 4 because we're still collecting sensor values over time. However, whereas Chapter 4 focused on *continuous motion*, this project works with *episodic motion*— gestures and movements that are performed only one or two times, not continuously. *Don't underestimate this difference* because it means a lot, as you'll see later.

The following are the steps to collect data from our board.

1. Read IMU (magnetometer data in particular).
2. Print CSV-encoded data to the Serial Monitor.
3. Collect data in Python as CSV files.

CHAPTER 5   TIME SERIES CLASSIFICATION USING PYTHON

Listing 5-1 contains the Arduino sketch.

Listing 5-1. Collect Magnetometer Data Sketch

```
/**
 * Listing 5-1: Read magnetometer data
 *
 * Required hardware: Arduino Nano BLE Sense.
 */
#include <Arduino_LSM9DS1.h>
#include <tinyml4all.h>

using tinyml4all::printCSV;

tinyml4all::LSM9DS1 imu;

void setup() {
  Serial.begin(115200);
  while (!Serial);
  Serial.println("Collect magnetometer data as CSV");

  // init IMU sensor (will throw an error on failure)
  imu.begin();
}

void loop() {
  // read magnetometer
  imu.readMagneticField();
  printCSV(millis(), imu.mx, imu.my, imu.mz);

  // no manual delay, default sample rate is ~80 Hz
}
```

You can debug the output by opening the Serial Plotter and bringing a magnet close to the board. You see the plot lines bending depending on the position of the magnet (see Figure 5-2).

CHAPTER 5 TIME SERIES CLASSIFICATION USING PYTHON

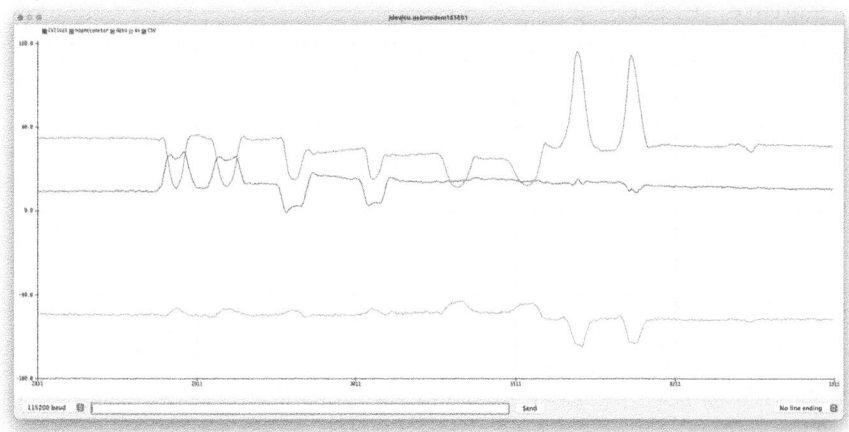

Figure 5-2. *Serial plot of magnetometer in the presence of a magnet*

To easily perform these actions comfortably, use tape to secure the magnet at the tip of your index finger.

After you confirm the sketch is working, create a new Python script. Listing 5-2 is almost a replica of Listing 4-2, but the names of columns were adjusted.

Listing 5-2. Capture Data from Serial in Python

```
from tinyml4all.time import capture_serial

while True:
    # prompt user for label and duration of sampling
    gesture = input("Which media control gesture is this? ")
    duration = input("How many seconds to capture? ")

    # exit when label or duration is empty
    if not gesture or not duration:
        break
```

175

CHAPTER 5   TIME SERIES CLASSIFICATION USING PYTHON

```
capture_serial(
    # * is a wildcard character that matches anything
    # on Windows it will look like COM1 or similar
    port="/dev/cu.usb*",
    baudrate=115200,
    # destination file
    save_to=f"media-control/{gesture}.csv",
    duration=f"{duration} seconds",
    # name of the columns
    headings="millis, mx, my, mz"
)
```

To run the code, open a terminal inside the folder where the script is located, activate the virtual environment, and type.

```
(tinyml)$ python capture_magnetometer.py
Press [Enter] when you're ready to start:
Which media control gesture is this? volume-up
How many seconds to capture?> 60
Task will start in 3...2...1...START!
Connected to serial port
100%|███████████████████████| 60/60 [00:60<00:00, 1.00s/it]
Collected 1085 lines of data
```

Caution   Always remember to record an *idle* class! Start the recording and do nothing, or move just a little bit.

Figure 5-3 shows some of the recorded gestures. Apart from the first (the *idle* class), you should be able to see the repetitions of each gesture a few times in each subplot.

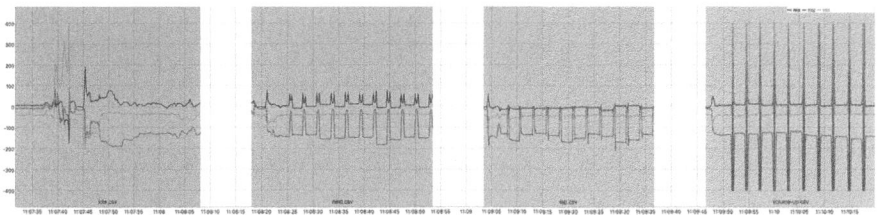

Figure 5-3. Time series plot of media control gestures

Data Labeling

While this step was "automatic" in Chapter 4, it now must be completed manually. This can be tedious if you have a lot of data, but remember that dataset preparation (collection and labeling) is usually time-consuming. Nevertheless, the higher the quality of your input data, the higher the quality of your classification results.

The tinyml4all Python package ships with a simple built-in time series labeling tool that runs in your browser. To start it, we must first load the time series data (Listing 5-3).

Listing 5-3. Load and Label Time Series Data

```
# note that the packaged changed from continuous
# to episodic!
from tinyml4all.time.episodic.classification import TimeSeries

ts = TimeSeries.read_csv_folder("media-control")

# run labeling GUI
ts.label_gui()
```

A new browser window opens with the plot of the time series and instructions on how to mark events of interest. In a nutshell, you only need to click in the middle of an event; a marker appears to confirm it. Repeat the process for each event, for each media control gesture (see Figure 5-4).

CHAPTER 5   TIME SERIES CLASSIFICATION USING PYTHON

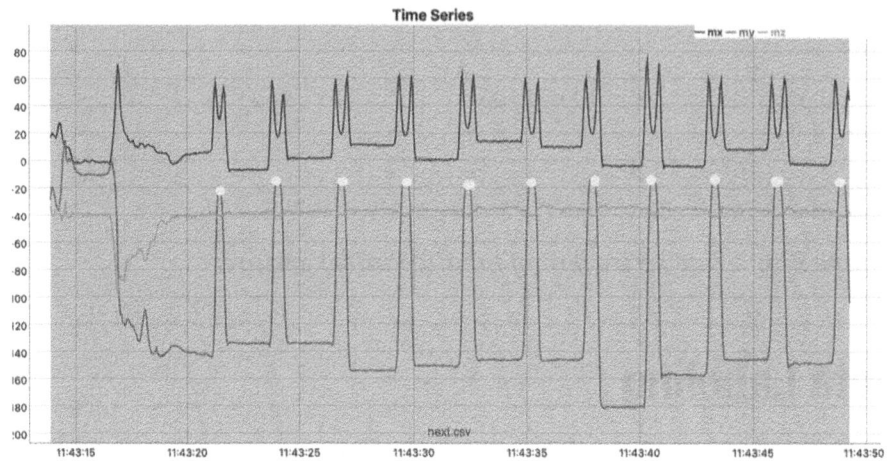

Figure 5-4. *Time series labeled events*

When you finish the labeling job, copy the contents from the Output tab (at the bottom of the page) and paste them into a new file called labels.json (or click the download icon) and move it inside the media-control folder. With this convention, the labels are automatically picked up from the TimeSeries.read_csv_folder function the next time you load your data.

Caution   You may find one or more PKL files inside your data folder. They are created automatically from the tinyml4all library and contain a compact representation of the time series data. Don't delete them since they allow faster loading time when you run your code more than once.

Confirm this by running the Python code shown in Listing 5-4.

Listing 5-4. Load Time Series Data with Labels and Events

```
from tinyml4all.time.episodic.classification import TimeSeries

ts = TimeSeries.read_csv_folder("media-control")
print(ts.events)
Events [volume-up (duration=800ms, count=24), volume-down
(duration=800ms, count=22), next (duration=800ms, count=22), back
(duration=800ms, count=23),) tap (duration=500ms, count=24)]
```

Feature Engineering

This chapter approaches the time series feature engineering manually. What makes a good feature in this context? As usual, we want to manipulate the input data so that the classifier at the end of the chain can easily classify it.

We often work with multidimensional data (made by many measurements): accelerometer, gyroscope, and magnetometer are made by three channels each; EMG (electromyography) data is usually made by three or eight channels; light sensors output R, G, and B values. Let's consider each measurement/dimension on its own; the features are extracted along each dimension, and values across dimensions are never mixed. For this reason, the rest of the chapter focuses on analyzing a single time series. In the case of multidimensional input data, we replicate the same analysis for each dimension and eventually concatenate the features of each to form a single feature vector.

CHAPTER 5   TIME SERIES CLASSIFICATION USING PYTHON

Note   Some tabular data feature engineering may also apply to time series data. Normalization, for example, is still a suggested operation that you should perform. Discretization and power transform (see Appendix A) have little significance in this new context instead.

Statistical Moments

Statistical moments are values that characterize the shape of a function. You should be familiar with them if you have some knowledge of statistics.

- **Mean**: The average value of a dataset. It's calculated by summing all values and dividing by the number of data points. The mean represents the central tendency of the data.

- **Standard deviation**: A measure of how spread out the data is from the mean. It's calculated by taking the square root of the average squared difference between each data point and the mean. A higher standard deviation indicates more variability in the data.

- **Skew**: A measure of the asymmetry of the data distribution. Positive skew means the tail of the distribution extends more to the right, while negative skew means it extends more to the left. A skew of zero indicates a symmetric distribution.

- **Kurtosis**: A measure of the *tail density* of the data distribution. It describes how heavy the tails are compared to a normal distribution. High kurtosis indicates more extreme outliers (many samples far from the mean), while low kurtosis suggests fewer outliers. A normal distribution has a kurtosis of 3.

This group includes the *minimum* and *maximum* values of the series.

Autocorrelation

Autocorrelation measures how similar a time series is to itself at different points in time. It tells us how much the current value in a time series is related to its past values. In autocorrelation, *lag* refers to the time difference between a current and past observations. For example, `lag=1` means comparing each data point with the one immediately before it, `lag=2` means comparing it with the point two steps back, and so on.

Autocorrelation can be positive (values tend to move in the same direction over time) or negative (values tend to alternate directions). The strength of autocorrelation ranges from –1 to 1, where 1 indicates perfect positive correlation, –1 indicates perfect negative correlation, and 0 indicates no correlation. It is a good feature descriptor because it captures short-term patterns, no matter the scale of the data (since it is based on a difference).

Shape Metrics

Finally, other descriptors can become useful for an accurate classification.

- **Number of peaks**: How many peaks (min or max values) are in the series? The higher this number, the less flat the series is.
- **Count above/below mean**: How many samples have a value greater/lower than the mean?

All these descriptors can be extracted from a single time series. If the time series covers a duration of hundreds of samples, these values may not capture its full underlying pattern. For this reason, let's break our long input segment into shorter, overlapping chunks using a *windowing* scheme.

Windowing

Windowing is a core operation of time series classification. It is the process of taking a long list of values and rearranging them into many sublists of fixed length. The time-domain features are then computed over each of these windows. This process generates a feature vector that is denser (captures patterns at a finer level of detail) and should help boost the accuracy of the classification. When you define a window, there are actually two parameters to configure.

- **Length**: How many samples are contained inside a window
- **Shift**: When the window is full, how many samples are discarded to make room for new ones

The shift parameter can be considered *rate limiting:* the shorter the shift, the more frequent the classification. This allows faster responses to changes in the data and serves well if you apply a moving average on the outputs (to avoid spurious misclassifications). In contrast, a large shift implies fewer CPU computations, leaving time to complete other tasks (if you have any).

To make for even *denser* features, it is also possible to chunk data into medium-sized windows (e.g., two seconds) and then chunk each window once more (e.g., 400 ms chunks). When dealing with complex input data, this approach should improve classification accuracy. In our project, the frequency of the magnetometer is 80 samples per second, and the duration of the events is under one second. Subchunking shouldn't make a huge difference, but for the sake of demonstration, Listing 5-5 uses a subchunk of 200 ms. If you set a subchunk duration, that becomes the shift of the window. Otherwise, you must specify this value manually.

One vs. Rest

Depending on your specific project, you may have many classes sharing all the same duration (the easiest case), or they may have different durations (the case no one talks about but that happens frequently in the real world). To accommodate all projects, the `tinyml4all` package treats all episodic time series classification tasks as a *one vs. rest* problem—even if all the labels share the same length.

If you recall from Chapter 1, one vs. rest means that many classifiers are trained, one for each class. The objective of each classifier is to learn to differentiate the class of interest from anything else (e.g., "volume up" vs. "not volume up" in our case). This introduces a lot of complexity behind the scenes, but it is mandatory to handle different use cases properly.

Episodic Time Series Classification Chain

Now that all the pieces are together, it is time to assemble a chain of operations that pre-process, rearrange into windows, extract features from, and classify our time series input data.

This time, the `Chain` class has a more structured format that mimics the different parts required.

- `pre` (for pre-processing): A list of steps that are applied to the input data no matter their label (e.g., scaling, normalization, binning). This is optional.

- `chunk`: Defines the duration of the subchunks. This is optional.

- `features`: A step (or list of steps) responsible for extracting features from each window (or chunk). The `tinyml4all` library ships with a few (as described earlier in this chapter), but you could implement your own if needed. This is mandatory.

- ovr: A list of steps to run after the feature extraction on binary views of the data (class of interest vs. not class of interest). Here, you put a classifier and any other step you see fit (e.g., feature selection). This is mandatory.

Listing 5-5 instantiates a chain with the following configuration.

- Min-max normalization
- Chunks of 200 ms
- Time-domain feature extraction
- Feature selection
- Random forest classifier

Listing 5-5. Episodic Time Series Classification Pipeline

```
from tinyml4all.time.episodic.classification import
TimeSeries, Chain
from tinyml4all.time.episodic.features import Window
from tinyml4all.time.features import Scale, Moments,
Autocorrelation, Peaks, CountAboveMean, Select
from tinyml4all.time.models import RandomForest

ts = TimeSeries.read_csv_folder("media-control")

chain = Chain(
    pre=[Scale("minmax")],
    window=Window(chunk="250ms", features=[Moments(),
    Autocorrelation(), Peaks(), CountAboveMean()],
    ovr=[
        Select(sequential="auto"),
        RandomForest()
    ]
)
```

```
tables = chain(ts)

# one vs rest classification produces N binary tables
# with classes "<class of interest>" and
# "not <class of interest>"
for table in tables:
    print(table.classification_report())
```

```
                 precision    recall  f1-score   support

not volume-up         1.00      1.00      1.00       230
    volume-up         1.00      1.00      1.00        26

     accuracy                             1.00       256
    macro avg         1.00      1.00      1.00       256
 weighted avg         1.00      1.00      1.00       156

+-------------------+---------------+-----------+
| True vs Predicted | not volume-up | volume-up |
+-------------------+---------------+-----------+
|     not volume-up |      230      |     0     |
|         volume-up |       0       |     26    |
+-------------------+---------------+-----------+
```

Wait a minute, something looks out of place! We were working with time series, so why does the `chain` call returns tables? We lost the time dimension because we extracted a list of features from each window of data that describes an entire chunk of values with a handful of metrics. Time was only needed to group samples together in windows based on timestamps. After the feature extraction, it has no more meaningful purpose.

CHAPTER 5   TIME SERIES CLASSIFICATION USING PYTHON

> **Note** When working with "traditional" machine learning models, you always want to reframe your data to look like a table because that's the format those models are meant to process. Chapter 8 explains that you may create neural network architectures that work straightaway with temporal data instead.

Deploy to Arduino

At this point, the training workflow is done. We implemented feature engineering and classification on our powerful desktop PC; the accuracy achieved by the classifier meets our expectations. Here is where the *tiny* part comes in. The last step is to finally convert our chain to C++ so that we can import the code into our embedded project. Listing 5-6 shows how to generate such code.

Listing 5-6. Convert Time Series Classification Pipeline from Python to C++

```
chain.convert_to("c++", class_name="MediaControlChain",
save_to="MediaControlChain.h")
```

Copy the generated file inside your project's folder to run the classification in your sketch. Then, upload the code in Listing 5-7. This sketch uses the USB HID features of the Arduino Nano BLE Sense, which allows the board to act as a keyboard connected to your PC to simulate key presses. If you're using a different board that doesn't support this protocol, you must devise a different actuation logic (e.g., send commands over Serial and write a script that runs on your PC to decode or send them over BLE).

CHAPTER 5 TIME SERIES CLASSIFICATION USING PYTHON

Listing 5-7. Arduino Sketch to Control Media Using Gestures

```
/**
 * Listing 5-7: Predict media control gestures
 *               using time domain features
 *
 * Required hardware: Arduino Nano BLE Sense
 */
#include <Arduino_LSM9DS1.h>
#include <PluggableUSBHID.h>
#include <USBKeyboard.h>
#include <tinyml4all.h>
// this is the file generated earlier
#include "./MediaControlChain.h"

tinyml4all::LSM9DS1 imu;
tinyml4all::MediaControlChain chain;
USBKeyboard keyboard;

void setup() {
  Serial.begin(115200);
  while (!Serial);
  Serial.println("Predict media control gestures");

  // init IMU sensor
  imu.begin();
}

void loop() {
  // read magnetometer
  imu.readMagneticField();
```

```
// try to run prediction
// if window is not full, will return false
if (!chain(imu.mx, imu.my, imu.mz))
  return;

// here the chain run successfully
// chain.output.classification.idx holds the most probable
    class index
// chain.output.classification.label holds the most probable
    class label

// ignore idle class
if (chain.output.classification.label == "idle")
  return;

// print the inferred gesture
Serial.print("Predicted ");
Serial.print(chain.output.classification.label);
Serial.print(" with confidence ");
Serial.println(chain.output.classification.confidence);

// trigger key presses based on label
if (chain.label == "tap")
  keyboard.media_control(KEY_PLAY_PAUSE);
else if (chain.label == "next")
  keyboard.media_control(KEY_NEXT_TRACK);
// and so on...

// for the list of available keys, visit
// https://github.com/arduino/ArduinoCore-mbed/blob/main/
    libraries/USBHID/src/USBKeyboard.h
}
```

CHAPTER 5   TIME SERIES CLASSIFICATION USING PYTHON

Open the Serial Monitor and start performing the media control gestures. You should see something similar to the output in Figure 5-5.

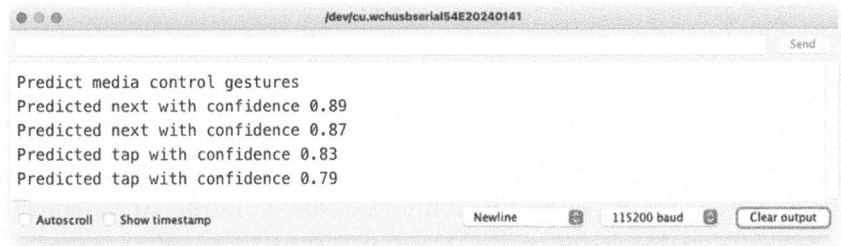

Figure 5-5. Media control gestures prediction output

The prediction time is about 4 ms. Comparing that with the 90 ms of the Edge Impulse model, you can appreciate how much faster this method is (the downside is that it is generally less accurate).

Tip   Refer to the online materials repository for a demo video of the project in action.

Summary

This chapter tackled the time series classification task by leveraging a code-oriented workflow that leverages manual episodic labeling and time-domain features. With respect to the low-code, Fourier transform–based approach, this requires a bit more configuration from our side, but the result is more lightweight and faster.

Accuracy greatly depends on the data you collect for training and the types of events you want to classify. Continuous motion is easier to detect thanks to its long duration and repetitive pattern. Episodic, one-time events require a more granular and reactive model to quickly grasp changes in almost real time. The chain introduced here tries to achieve such results.

The next chapter introduces a new data type: audio. While *technically* still a time series (air pressure variation over time), in practice, audio deserves a tailor-made processing framework due to its high frequency (8 or 16 kHz, usually) and specific interpretation by the human ear/brain that goes beyond the pure temporal analysis.

CHAPTER 6

Audio Wake Word Detection with Edge Impulse

In the world of TinyML and audio processing, audio wake word detection—a.k.a. *keyword spotting* (KWS)—has become essential for enabling hands-free interactions with our devices. From smart speakers to mobile phones, KWS systems constantly listen for specific wake words or commands. However, implementing these systems on resource-constrained devices presents unique challenges, particularly when processing raw audio data efficiently.

So far, you worked with numeric data that represented sensor values. Those values could be well arranged into a tabular format, either with a timestamp (time series) or without one. A completely different type of data is audio. Even if, from a technical point of view, audio is still a time series, its intrinsic meaning goes far beyond its mere waveform. The human brain can associate much more information with audio signals than regular time series, and we can detect sounds and words from them.

CHAPTER 6   AUDIO WAKE WORD DETECTION WITH EDGE IMPULSE

This chapter is devoted to audio classification using the Edge Impulse low code platform. You take a brief tour of how audio can be captured using a digital microcontroller, which features constitute the state-of-the-art descriptors for speech analysis, and how to detect specific words from your board. As a reference project for this chapter, we build a clone of the widely used voice assistant devices (Alexa, Siri, Google Home), which are limited to their activation mechanism. Instead of "OK Google," you can wake up your Arduino board with the "Hey Arduino" vocal command (see Figure 6-1).

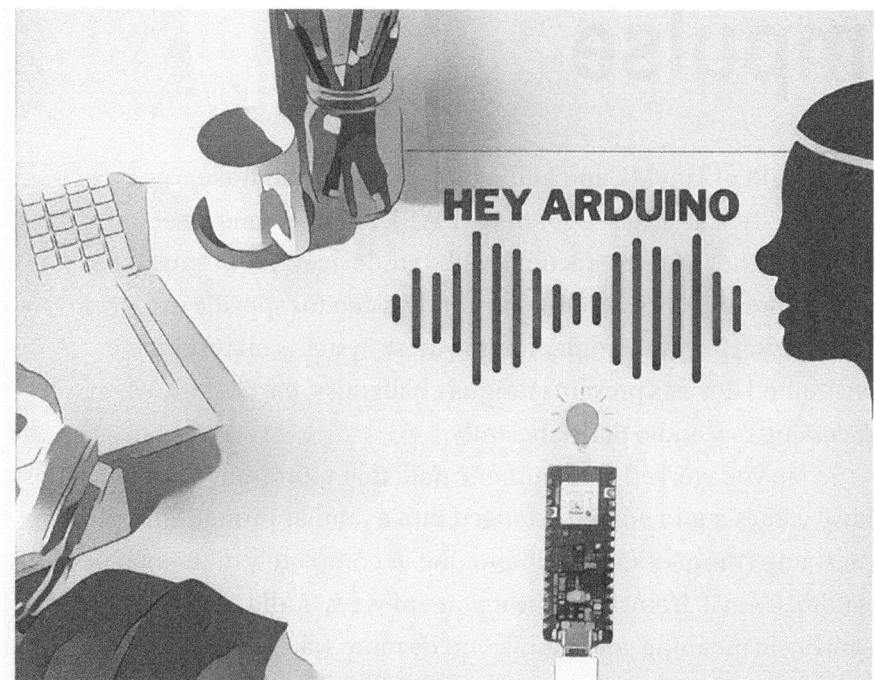

Figure 6-1. Audio wake word detection

Hardware Requirements

You need a microphone to capture audio from your Arduino board. A few boards come with a microphone built-in, including the following.

- Arduino Nano BLE Sense, Arduino RP2040 Connect, and Nicla Vision (PDM interface)
- Seeed Studio Wio Terminal and LILYGO TTGO variants (I2S interface)

If you don't have any of these, you can buy an external microphone and hook it up according to its requirements. Usually, mics for Arduino boards come in two shapes.

- **PDM (pulse-density modulation)**: The audio waveform is encoded in the density of the digital pulses and requires specialized software to be decoded. It usually only handles mono audio with 16 bits per sample.
- **I2S (Inter-IC Sound)**: The audio output comes already encoded, so there's no need for further processing. I2S mics in the Arduino ecosystem can usually handle stereo audio at 32 bits per sample.

Check that the mic you buy is compatible with your board before you order!

The code listings in the rest of this chapter assume you are using an Arduino Nano BLE Sense. You can find more examples for different devices in the book code repository.

Software Requirements

Depending on the type of microphone you're going to use, you need the respective Arduino library. The Arduino boards I mentioned ship with an official PDM.h library from the Arduino core itself. Boards from other vendors may ship with similar libraries. When using the I2S mics, you'll usually leverage the built-in I2S.h library.

The tinyml4all Arduino library comes with some utility adapters for most common configurations to zero the differences between models, but 100% coverage is not guaranteed.

On the Python side, you only need the tinyml4all package (which you should have already installed).

Capture Data

To form our dataset for the "Hey Arduino" wake word detection, we will speak those words aloud at least 30 times while recording them with our board. These samples make our *positive* dataset. As highlighted in Chapter 1, a classification task requires all the classes to be known beforehand. But we only have a single class here, right?

Not really.

It turns out that when dealing with problems with only a single *class of interest*, you can reframe the problem as a binary classification task. The two classes are the class of interest ("Hey Arduino") and the *not* class of interest/unknown (everything but "Hey Arduino").

This setting is important, and you need to be aware of it. During training, your classifier should experience the whole spectrum of possible inputs to be reliable. If you only show it a limited set of cases (e.g., "Hey Arduino" and silence), how can you expect it to respond correctly to an unknown sound (e.g., "Thank you")?

CHAPTER 6 AUDIO WAKE WORD DETECTION WITH EDGE IMPULSE

Many tutorials on the web fail at this point: they tell you to record the wake word and silence only. If you were to follow that path, chances are that the model would respond to the wake word and pretty much every other sound since that would be more similar to the wake word than it would be to silence. If you don't want to spend a long time gathering data by yourself, a reliable approach is to download a "sounds in the wild" dataset that contains pre-recorded audio samples of different sounds (environmental, synthetic, or human voice; e.g., traffic jam, dishwasher, ringtones, animal roars), as highlighted later.

Tip Since this is the era of accessible generative AI, I'll also show you how you can leverage text-to-speech synthesis to generate artificial training data to make your model better generalize to voices different from yours.

That considered, after the 30 positive samples, let's collect at least the same number of *negative* samples. They can be silence, noise, songs from Spotify, or us saying something different from the wake word. What matters the most is that these samples should cover as much as possible the spectrum of inputs our board could receive once it has been deployed and is ready to wait for the designated word to be pronounced.

Note Even if not strictly necessary, we will set up a classification task with three classes: *wake word, unknown* (spoken words), and *noise* (environmental sounds). This should help the model better cluster similar samples into their specific class instead of merging everything under the same label.

CHAPTER 6 AUDIO WAKE WORD DETECTION WITH EDGE IMPULSE

Audio Data Format

Audio data comes at a very high speed. For example, the built-in mic from the Arduino Nano BLE Sense has a sampling frequency of 16 kHz. That means that every second, you must read, store, and transmit 16,000 values! Since the generated values are 16-bit signed integers (from –32768 to 32767), the most compact way to transmit them from the board to the PC is the binary format. If we were to adopt the ASCII encoding (writing them as plain numbers) with CSV format, we could end up using up to six characters per value (five for the numbers, one for the comma): that would be a waste of space.

This means the copy-paste approach to data capture is ruled out. Since the Arduino Nano BLE Sense doesn't have Wi-Fi hardware or a built-in SD card reader, we will ignore them, too. (You can still attach an external SD card reader if you want.)

The only one remaining is the ingestion from the `tinyml4all` Python library. This time, we won't create a single file per class with all the data packed inside. We want a single file for each audio sample to manually check whether it's good or not and easily handle it throughout our desktop filesystem.

Edge Impulse also has a built-in data capture mechanism integrated into the browser. It is particularly handy to capture data from your smartphone but to use your microcontroller, you must install the Arduino CLI (command-line interface), and unless your browser supports the WebUSB standard, you must install their own CLI. We won't go this route, but you can browse their documentation to learn more.

Arduino Sketch

This section varies based on your board and microphone. Listing 6-1 is for the Arduino Nano BLE Sense with a built-in PDM microphone. You can find more examples of other variants in the book code repository.

The sketch is provided to be run as is, without modifications. The only parameter you may be interested in changing is the volume. It accepts a value from 0 (minimum) to 100 (maximum). The higher the volume, the stronger the intensity of your voice when recorded.

Caution You start hearing distortions at high volumes (70 and above), so don't exaggerate.

Listing 6-1. Arduino Sketch to Capture Audio from PDM Microphone

```
/**
 * Listing 6-1: Collect audio data from PDM microphone.
 *
 * Required hardware: Arduino Nano BLE Sense
 * Required hardware: or any board with a PDM microphone
 *                    (e.g., Nano RP2040 Connect)
 */
#include <PDM.h>
#include <tinyml4all.h>

tinyml4all::PDMicrophone mic;

void setup() {
  // increase Serial speed
  Serial.begin(115200 * 2);
  while (!Serial);
  Serial.println("Collect audio data");
  // refer to the board datasheet for the mic frequency
  // (Nano RP2040 Connect frequency is 21kHz, for example)
  mic.frequency("16 khz");
```

CHAPTER 6 AUDIO WAKE WORD DETECTION WITH EDGE IMPULSE

```
  // volume goes from 0 to 100
  mic.volume(30);

  // configure microphone
  mic.begin();
}
void loop() {
  // await data to be available
  mic.await();

  // send over Serial
  mic.print();
}
```

After the sketch is flashed, confirm that the board is streaming the audio data over Serial by opening the Serial Monitor. The output will look like garbage since it's binary data, not text (see Figure 6-2). The important thing is that something gets printed. Now, it's time to run the Python collector script.

Figure 6-2. *Audio binary serial output*

CHAPTER 6 AUDIO WAKE WORD DETECTION WITH EDGE IMPULSE

Python Code

You must configure a few parameters before running the Python script (Listing 6-2).

- The number of samples (not seconds!) you want to capture
- To which port the board is connected
- Where to store the audio files
- How long the wake word lasts
- The frequency of the captured audio

Caution Double-check that the frequency in Python matches that of your microphone. Arduino Nano BLE Sense has a 16 kHz sample frequency, and Arduino Nano Rp2040 Connect has a 21 kHz frequency instead, for example.

Double-check that the baud rate in Python matches that of the Arduino sketch!

Listing 6-2. Capture Wake Words from Serial in Python

```
from tinyml4all.audio import capture_serial

if __name__ == '__main__':
    capture_serial(
        # * is a wildcard match
        port="dev/cu.usb*",
        baudrate=115200 * 2,
```

```
    num_samples=30,
    word_duration="2 seconds",
    save_to="wakeword",
    mic_frequency="16 khz"
)
```

The program prompts you to confirm the start of the capture of each sample by pressing [Enter]. After you confirm, say your word aloud. A log message lets you know the status of the capture.

The code to capture non-wake words is the same (see Listing 6-3). Just save the files to a different folder.

Listing 6-3. Capture Non-Wake Words from Serial in Python

```
from tinyml4all.audio import capture_serial

if __name__ == '__main__':
    capture_serial(
        port="dev/cu.usb*",
        baudrate=115200 * 2,
        num_samples=30,
        word_duration="2 s",
        save_to="unknown",
        mic_frequency="16 khz"
    )
```

Run both the scripts. You should have two folders with 30 audio files inside each. The audio capture process may not be perfect, so manually listen to each sample to confirm it is good. If not, delete the file and re-run the script.

Third-Party Datasets

Collecting data on your own usually requires a lot of time and effort. Even after you collected a handful of data, you run the risk that the model may not generalize well. In the context of our keyword spotting project, let's say you repeated 100 times the wake word you want to recognize. This seems like a legitimate amount of data to train a model, right? Yes, but only if you can tolerate the model only recognizing your voice!

Generalization is the ability of a model to recognize the same wake word pronounced by people other than you. This is how commercial home assistants work since they're intended for the mass market. If you want to achieve a similar level of generalization, you need more data in both quantity and variety.

A quick and easy shortcut to this is searching online for an existing dataset. The data science industry is helpful when sharing their data, so something related to your current project may already exist on the web, such as Common Voice from Mozilla (`https://commonvoice.mozilla.org/en`). This greatly speeds up your data collection process and allows you to add a high degree of variability to your data.

Edge Impulse provides a dataset for the keyword spotting problem [1]. It contains hundreds of samples for the *Yes, No, Unknown* (any word that is neither *yes* nor *no)* and *Noise* (environmental sounds) classes. We're not interested in the *Yes* and *No* files since we already have our own wake word, but we'll borrow the other two classes of files. So navigate to the dataset page, download the ZIP file and extract it on your computer.

Synthetic Wake Word Generation

Artificial intelligence has become ubiquitous in the last few years. With the exploitation of large language models, the industry of generative artificial intelligence is experiencing highly accelerated growth in the science landscape. It turns out that we can leverage generative models for our TinyML project, too.

Thanks to mainstream text-to-speech synthesis tools readily available today, we can quickly generate many variations of tone, pronunciation, and pitch in seconds. I've made a dataset publicly available with tens of samples for our "Hey Arduino" wake word that you can download from the book repository.

Next, let's discuss how to generate such a dataset.

Azure Text-to-Speech

Many platforms offer text-to-speech (a.k.a. TTS) tools. Most are paid services, but the largest ones—like Microsoft Azure and Amazon Web Services—have a free plan available. The `tinyml4all` Python package provides an implementation based on the Azure TTS service, so you need to register and create a free cloud instance on their website [2].

Once done, you need two pieces of information from your Azure dashboard (see Figure 6-3).

- **api_key**: An alphanumeric string that identifies you (either *key1* or *key2*).

- **region**: The data center your cloud instance is deployed to.

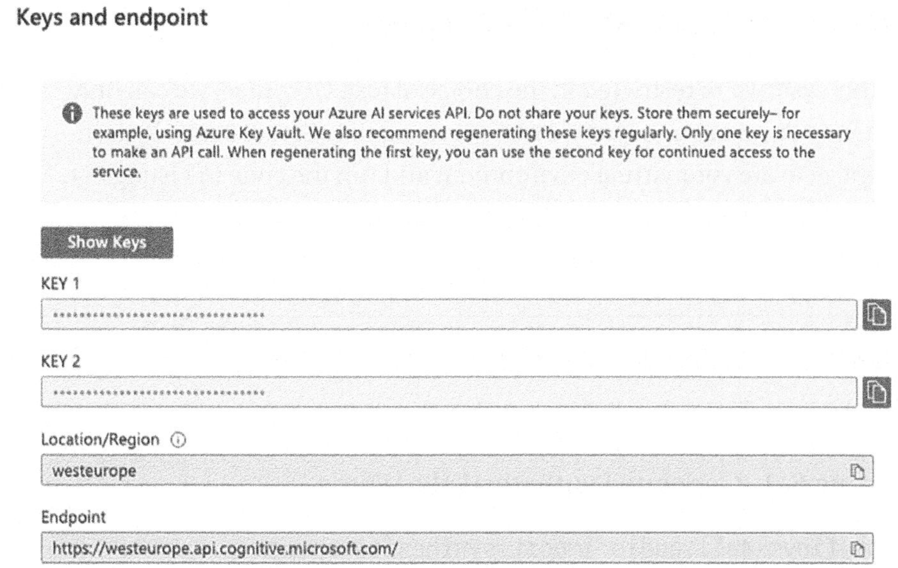

Keys and endpoint

> These keys are used to access your Azure AI services API. Do not share your keys. Store them securely– for example, using Azure Key Vault. We also recommend regenerating these keys regularly. Only one key is necessary to make an API call. When regenerating the first key, you can use the second key for continued access to the service.

Show Keys

KEY 1
`********************************`

KEY 2
`********************************`

Location/Region
`westeurope`

Endpoint
`https://westeurope.api.cognitive.microsoft.com/`

***Figure 6-3.** Azure dashboard credentials*

Now, you can generate two distinct datasets in seconds: *wake word* and *unknown*.

The dataset generation process is articulated in the following steps.

1. For the input language, download the list of available synthetic voices. The number of voices may range from as much as 56 for English to as low as 2 (male and female) for minor languages. Using varying voices makes the model robust to different people's tones and accents.

2. Generate combinatorial tuples of {voice, pitch, speech rate, volume} to increase variability even further.

3. Synthesize speech from text with each of the combinations.

For the English language, this can quickly add up to 500+ samples generated for a given text!

CHAPTER 6   AUDIO WAKE WORD DETECTION WITH EDGE IMPULSE

How to Generate a Wake Word Dataset

In this case, we're restricted in the choice of text. Of course, we can only generate the exact keyword we want to recognize. Create a new Python script, activate your virtual environment and run the code in Listing 6-4. When the script finishes running, you will find 30 audio files in the specified folder (it should take less than one minute to complete).

Caution   Replace the values in bold with your own!

Listing 6-4. Generate Synthetic Wake Words

```
from tinyml4all.audio import synthesize_speech

synthesize_speech(
    api_key="<your_azure_key>",
    region="<your_azure_region>",
    language="en-US",
    save_to="synthetic/wakeword",
    text="Hey Arduino",
    # must match with the duration in Listing 6-2
    duration="2 s",
    freq="16 khz",
    num_samples=30,
    # pitch percent variations
    # negative means lower pitch
    pitches=[0, -15, 15],
    # rate percent variations
    # negative means slower
    rates=[0, -10]
)
```

CHAPTER 6   AUDIO WAKE WORD DETECTION WITH EDGE IMPULSE

> **Caution**   *duration* and *freq* must match the ones you used while capturing real-world data using your Arduino board!

How to Generate an Unknown Dataset

In *unknown* words, there is more freedom. Text-to-speech can only synthesize spoken words, not environmental sounds, so let's choose a random selection of words that share a similar duration with our wake word. Feel free to input as many variations as you prefer so that the dataset covers a lot of real-world scenarios. The code in Listing 6-5 is nearly the same as Listing 6-4; the only parameters that change are the save_to folder and text.

Listing 6-5. Generate Synthetic Unknown Words

```
from tinyml4all.audio import synthesize_speech

synthesize_speech(
    api_key="<your_azure_key>",
    region="<your_azure_region>",
    language="en-US",
    save_to="synthetic/unknown",
    text=["Hello world", "Thank you", "Good morning"],
    # must match with the duration in Listing 6-2
    duration="2 s",
    freq="16 khz",
    num_samples=50,
    # pitch percent variations
    # negative means lower pitch
    pitches=[0, -15, 15],
```

CHAPTER 6   AUDIO WAKE WORD DETECTION WITH EDGE IMPULSE

```
    # rate percent variations
    # negative means slower
    rates=[0, -10]
)
```

Feel free to generate as much synthetic data as you see fit. I'd recommend at least 50 samples.

Caution   Audio labels are inferred from the innermost folder name of the files. So be sure that both recorded and synthetic samples share the same folder name (*wakeword* and *unknown*)!

Load and Inspect Data

Inspecting audio data visually in bulk is less intuitive than tabular and time series data. Even though a single audio sample can be plotted much like time series data (though with a single dimension—the intensity), plotting many samples one after the other suffers a couple drawbacks: the plot can become very long (16,000 samples per second × 30 samples × 2 seconds each makes nearly 1 million points!), and you wouldn't be able to easily compare samples that are not adjacent.

Keeping in mind that the best way to evaluate the quality of your samples is by listening to each, you can attempt to perform a visual inspection of how similar our wake words look by plotting them one on top of the other, aligned on the time axis. This way, each sample should resemble a common pattern, and outliers should be easy to spot. To make it easy to identify (possible) outliers, let's generate an interactive plot (see Figure 6-4) that opens in your web browser and highlights the sample under your mouse cursor (see Listing 6-6).

Listing 6-6. Plot Audio Files from a Folder (Overlapping)

```
from tinyml4all.audio import Album

album = Album.read_wav_folders(
  "wakeword",
  "unknown",
  "synthetic/wakeword", "synthetic/unknown"
)
# plot samples (overlapping)
album.overlap_plot(
  palette="magma",
  samples_per_class=50,
  points_per_sample=1_000)
```

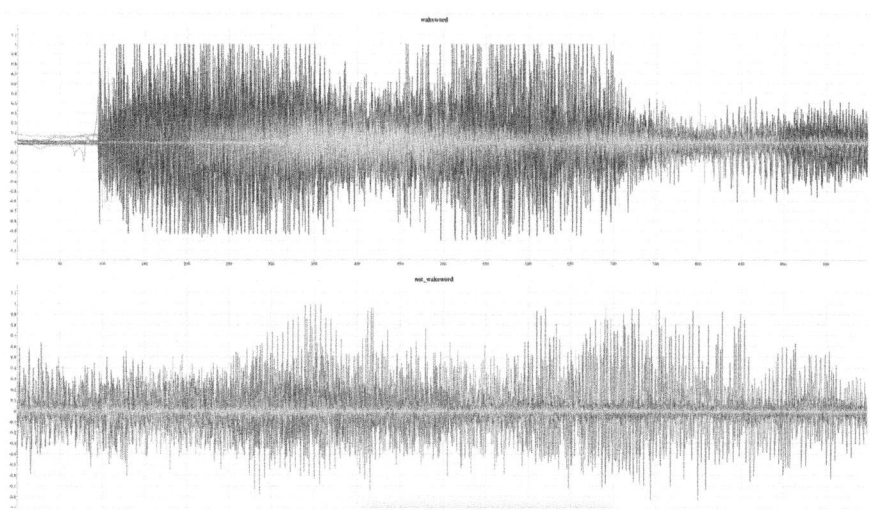

Figure 6-4. *Overlap plot of audio samples*

If you still want to plot each sample on a single plot, one after the other, refer to Listing 6-7. The result is displayed in Figure 6-5.

Listing 6-7. Plot Audio Files from a Folder (Sequential)

```
album.sequential_plot(
  palette="viridis",
  samples_per_class=10,
  points_per_sample=1_000)
```

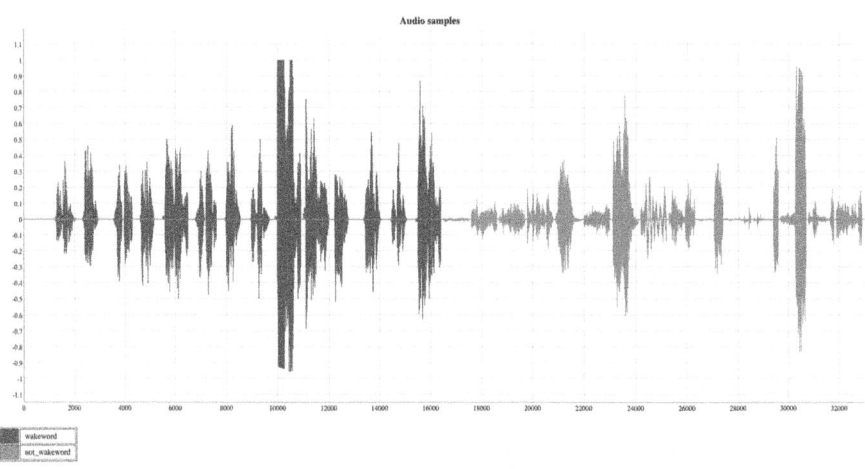

Figure 6-5. *Sequential plot of audio samples*

In Figure 6-4, all the audio samples of the two classes are plotted aligned on the same time axis. Even though the absolute value of the intensity may vary slightly from sample to sample, a recurring pattern is clearly visible. There may be a bit of misalignment every now and then, but that's totally expected. Move your mouse cursor over the plot to highlight a single sample.

In Figure 6-5, wake words are colored in blue, and unknowns in green. Samples are spaced apart from each other for improved readability. You should be able to see that wake words share a similar shape (even if there are differences between recorded and synthetic samples). The *unknown* samples differ from each other. (The first few samples are silence, thus the low intensity.)

Our data looks good. There are no clear signs of malformed samples. If that's not the case with your dataset, delete the corrupted or plain wrong files and collect a few more good samples.

Edge Impulse Data Acquisition

After you have collected a good amount of quality data, move on to https://edgeimpulse.com and create a new project named *keyword_ spotting*. Like in Chapter 4, you must upload your files. This time, there's no wizard to configure beforehand, and you can immediately go to the upload form. Upload data in three sessions: first, upload the *wake word* samples, then the *unknown*, and finally, the *noise* (from the Edge Impulse synthetic dataset downloaded earlier). Each time, be sure to do the following.

1. Enter the correct label manually.

2. Tick the **Automatically split between training and test testing** checkbox (see Figure 6-6).

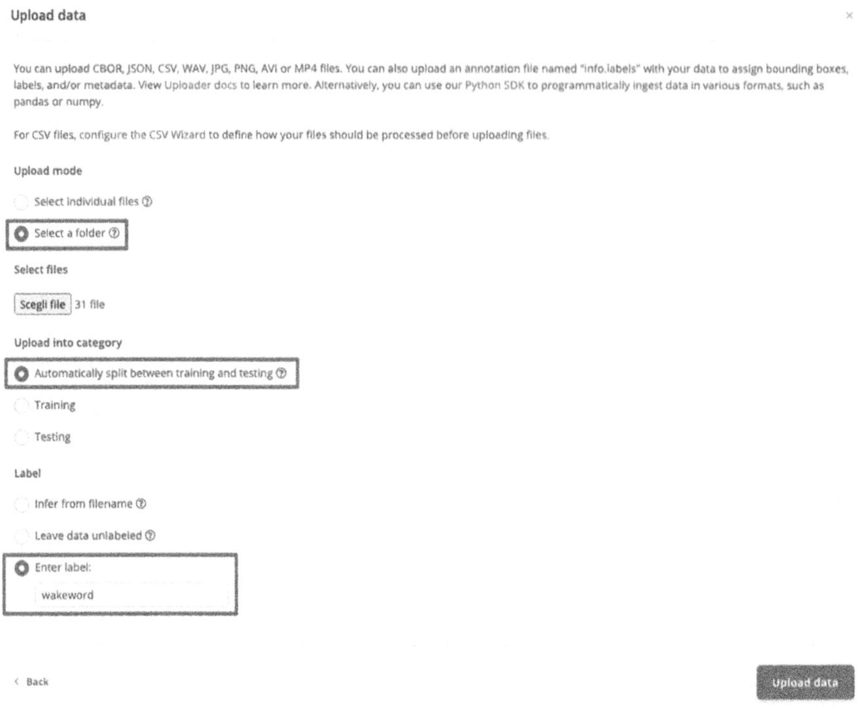

Figure 6-6. Edge Impulse data upload form

Note Upload both your own data and synthetic/downloaded data!

Feature Engineering

At this point in the book, you should know that feature engineering is a crucial step in a machine learning pipeline. Audio data makes no difference. Besides the usual benefits (better model learning), additional benefits are specific to this data type.

- **Dimensionality reduction and computational efficiency**: Raw audio data contains a vast amount of information, much of which is not relevant for keyword detection. Feature extraction helps reduce this dimensionality, focusing on the most important aspects of the signal, thus creating a more compact and computationally efficient representation.

- **Noise reduction**: By extracting specific features, we can often separate the signal of interest from background noise, improving the robustness of our KWS system.

Several considerations must be made to accommodate the human perception of sound is not linear. The standard feature descriptors for keyword spotting are called *Mel-frequency cepstral coefficients* (MFCCs), but before you can understand what they are and how they're computed, you need to know what a *Mel spectrogram* is.

Mel Spectrogram

The Mel spectrogram is a visual representation of sound that considers how humans perceive different frequencies. It is widely used in audio analysis and speech recognition tasks. To understand the Mel spectrogram, let's break down its components and the process of its creation.

- **Short-time Fourier transform (STFT)**: The first step in creating a Mel spectrogram is to perform STFT on the audio signal. The STFT is a technique that analyzes how the frequency content of a signal changes over time. It does this by dividing the signal into short, overlapping segments (usually 20-40 ms each) and applying a Fourier transform to each segment. The Fourier transform converts the time-domain signal into the frequency domain, showing which frequencies and their respective amplitudes are present.

- **Spectrogram**: The result of the STFT is a spectrogram, a two-dimensional signal representation. The horizontal axis represents time, the vertical axis represents frequency, and the intensity at each point represents the amplitude or energy of that frequency at that particular time. For a more intuitive understanding, you can visualize this matrix as an image (see Figure 6-7, left).

- **Mel scale**: The human ear perceives pitch in a non-linear fashion. We are more sensitive to changes in lower frequencies than in higher frequencies. The Mel scale is a perceptual scale of pitches that aims to mimic this human perception. It is approximately linear below 1000 Hz and logarithmic above 1000 Hz.

- **Mel filter banks**: to convert the spectrogram to a Mel spectrogram, we apply a series of triangular filters called Mel filter banks. These filters are spaced according to the Mel scale, with more filters in the lower frequency range and fewer in the higher frequency range. Each filter computes the average energy in a specific frequency range, weighted by the filter's triangle shape.

- **Logarithmic compression (dB)**: After applying the Mel filter banks, we take the logarithm of the resulting energies. This step is crucial because human perception of loudness is approximately logarithmic. It helps compress the spectrogram's dynamic range, making quieter sounds more visible and reducing the dominance of louder sounds (see Figure 6-7, right).

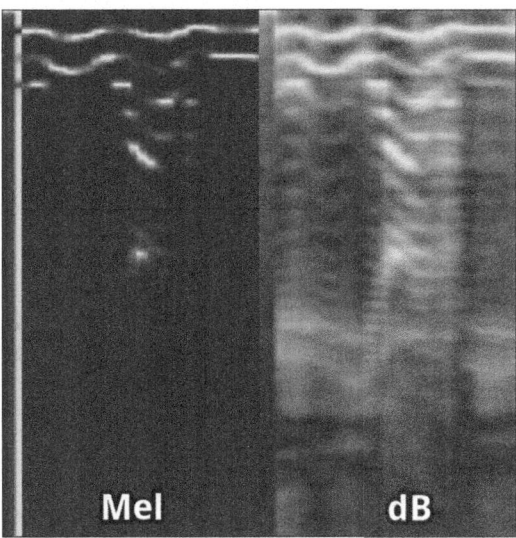

Figure 6-7. *Mel spectrogram before (left) and after (right) logarithmic compression*

- With respect to the raw STFT matrix, the vertical axis of the Mel spectrogram represents Mel frequency bands instead of linear frequency, and the color or intensity represents the log-compressed energy in each Mel band at each time point.

Why is the Mel spectrogram so effective for audio keyword spotting?

- **Perceptual alignment**: The Mel scale aligns well with human auditory perception, focusing on the frequency ranges most important for speech recognition.

- **Time-frequency representation**: Mel spectrograms capture both temporal and spectral information, which is crucial for identifying speech patterns.

- **Robustness**: Mel spectrograms are relatively robust to small variations in pitch and speaker characteristics.

Mel-Frequency Cepstral Coefficients

The Mel spectrogram is already a good candidate as a feature descriptor. It is a matrix that could be fed as input to a convolutional neural network (CNN) with good results out of the box. An MFCC is a process that generates a subset of that matrix (kind of feature selection). But if the Mel spectrogram is a valid option already, why are we adding another step? Is it really worth it?

Yes, it often is. MFCC extraction achieves two important objectives.

- **Effective dimensionality reduction**: Dimensionality reduction was introduced in Chapter 2. MFCCs, however, are computed in a different, audio-specific manner. They only select 12 to 20 features from the original Mel spectrogram using a *discrete cosine transform* (DCT), which generates highly decorrelated outputs.

- **Noise robustness improvement**: By discarding higher frequency components, MFCCs exhibit better rejection of noise.

The result of this process is a matrix of MFCC values, where each column represents a time frame, and each row corresponds to a cepstral coefficient. The number of rows is equal to the number of retained coefficients (12–20), and the number of columns depends on the duration of the audio and the frame rate used in the analysis (e.g., 1 second of audio divided into non-overlapping windows of 40 ms = 25 slices of time).

CHAPTER 6 AUDIO WAKE WORD DETECTION WITH EDGE IMPULSE

Audio Classification with Edge Impulse

Being a low-code tool, Edge Impulse hides all the implementation details of MFCCs from you. On the Create Impulse page, make the following selections.

- **Window size**: It should match the audio samples (2000 ms).

- **Window increase**: It should be large enough to allow the inference to run but not too large so as not to miss keywords. Also, it should be an integer divisor of 1000 to align with the `tinyml4all` Arduino library. A good default is 250 ms.

- **Processing block**: Audio (MFCC).

- **Learning block**: Classification.

Next, move to the MFCC page on the side menu to configure the feature extractor (see Figure 6-8). You can tweak a few options, but you should know what you're doing before making any changes. The defaults work well in our case, so we won't touch this page. Feel free to come back here if you later find out that your model is not performing well. The following are some options for tuning.

- **Number of coefficients**: You can increase this value to increase the number of features retained.

- **Frame length and stride**: You can increase the frame and make the stride a divisor to create overlapping windows.

- **FFT length**: Larger values compute more coefficients. Since FFT is a costly operation, I suggest you keep this value as low as possible.

Save the parameters and generate features on the next tab; when the process finishes, the feature explorer can give you an idea about how well the features are clustered.

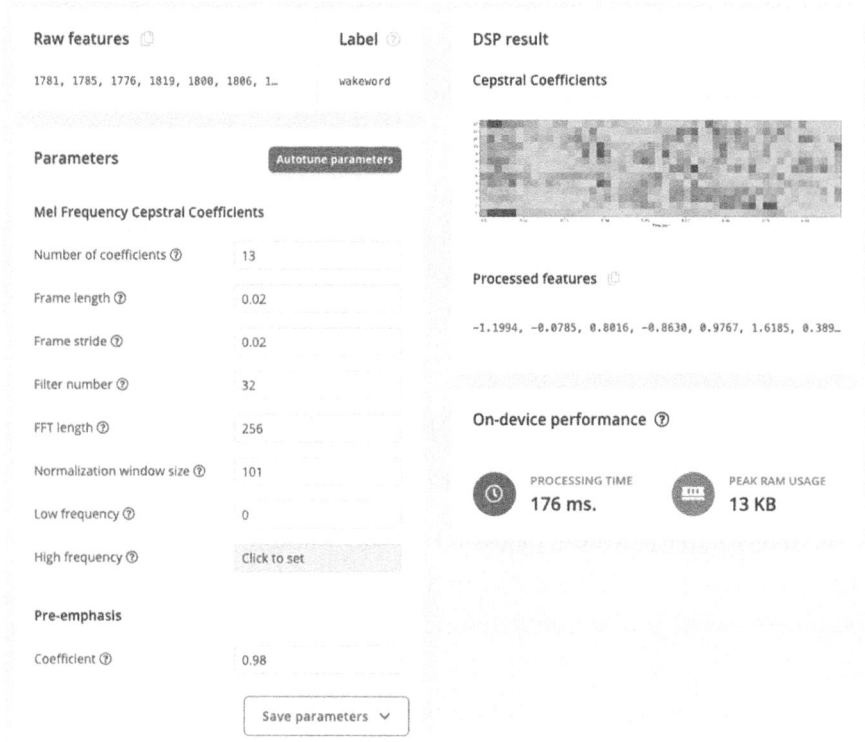

Figure 6-8. Edge Impulse MFCC configuration

Convolutional Neural Networks

We're now at the classification stage. Chapter 4 used a fully connected neural network made of two layers of neurons. Now that our features are a list of vectors, that topology may not be the best solution. A different topology often achieves state-of-the-art accuracy for image classification tasks: convolutional neural networks (CNNs)

CHAPTER 6   AUDIO WAKE WORD DETECTION WITH EDGE IMPULSE

Chapter 8 digs deeper into CNNs. At this stage, you only need to know that convolution may happen at different dimensional depths: 1D and 2D are the most common. 1D convolution works on sequential data (e.g., raw time series), while 2D convolution works on matrixes (e.g., images).

Convolution is an efficient operator that considers small regions of data at a single time (called *patches*) instead of each value individually—as fully connected networks do. A patch can either be a short sequence in the 1D case or a small matrix in 2D (common values for image classification tasks are 3×3 or 5×5, for example). The convolution *filter* (also known as the *kernel*, a matrix of coefficients of the same size as the patch) slides over each region of the input data and performs the dot product between the two (that is, the sum of element-wise multiplication of its coefficients with the values of the patch) to produce a single output value (see Figure 6-9).

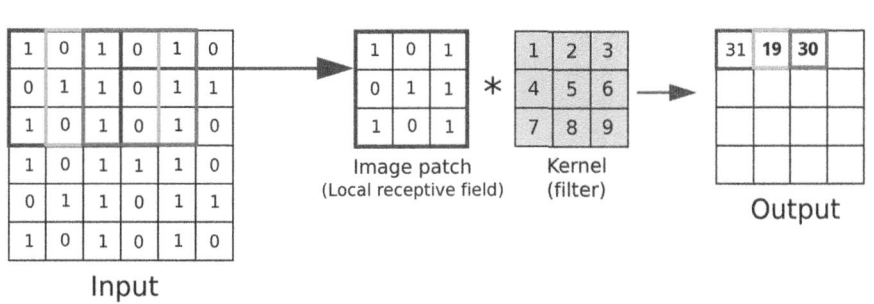

Figure 6-9. 2D convolution procedure

The kernel coefficients are initialized with random values; the learning process then consists of updating their values so that the classification results yield the highest accuracy.

From a memory point of view, convolution exhibits a huge improvement over full connection because the number of weights to store is independent of the input size. For example, the same 3×3 convolution filter works on 96×96 images as on 512×512 images; it just needs to slide more times.

CHAPTER 6 AUDIO WAKE WORD DETECTION WITH EDGE IMPULSE

MFCCs are often represented as images, but technically, they aren't (they don't depict real-world entities, and the vectors are not really "spatially" correlated). This is why Edge Impulse lets you choose which architecture you prefer for audio classification. 1D convolution is the most natural choice for this type of input data, but you can still treat the MFCC matrix as if it were an image and use 2D convolution. Since 1D convolution is more lightweight than 2D and generally exhibits better accuracy, it is selected by default (see Figure 6-10).

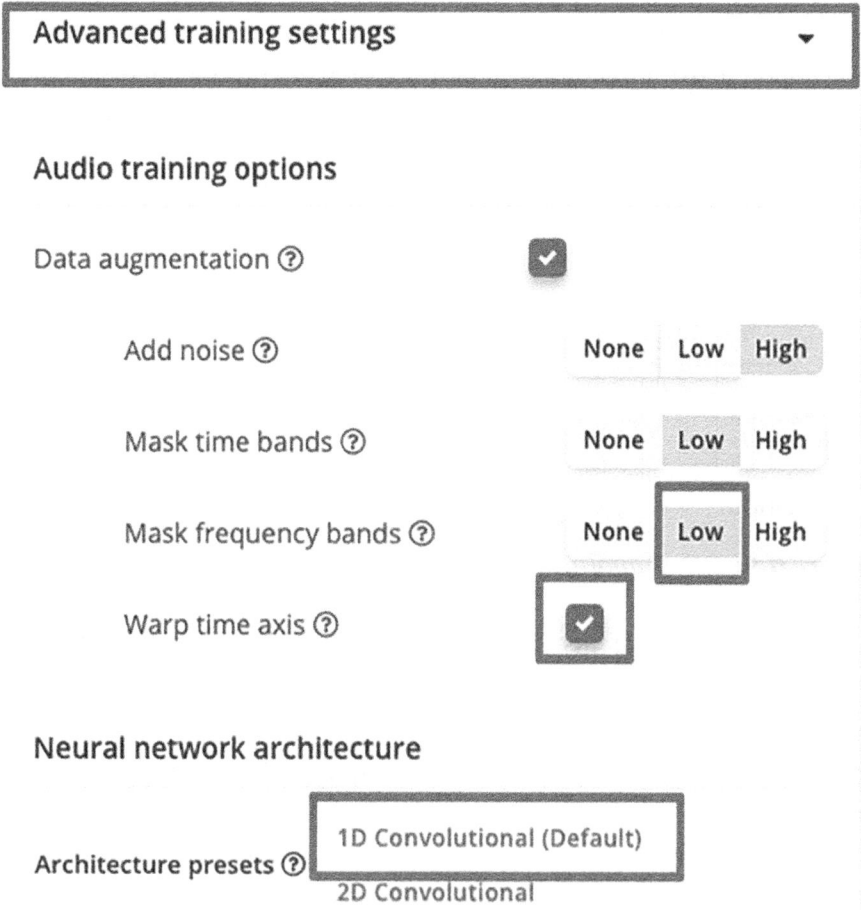

Figure 6-10. Edge Impulse training configuration

You can also tweak a couple of training parameters for faster training.

- **Number of training cycles**: This addresses how long the model should learn. With few samples, 50 is a good default.

- **Learning rate**: It controls how much the model should update its weights at each epoch. My go-to value is 0.005 (lower values may result in higher accuracy but may require more training cycles).

Since our dataset is limited, we can train the model for as few as 30–50 epochs while still achieving very good accuracy, as shown in Figure 6-11. When starting out and iterating fast on your model, keep the epochs number low and monitor if the metrics displayed in the logs improve over time. It is a waste of time to wait for 100 epochs of training if the model accuracy stays the same from epoch 30 (which happens frequently with easy datasets).

If your training goes like mine, you may be worried about the "low" accuracy on *noise* and *unknown* (about 85%). But remember that this project is only interested in the wake word class. If the model confuses a sample of the unknown class with one from the noise class, it is not a problem for us!

CHAPTER 6   AUDIO WAKE WORD DETECTION WITH EDGE IMPULSE

Figure 6-11. Edge Impulse confusion matrix

Testing

If the confusion matrix on the training page satisfies your expectations, move to the Model Testing page (from the menu on the left) to assess the model accuracy on audio it has never seen during training or validation. Click the **Classify all** button and wait for the results. In the best scenario, all the rows are green (indicating a perfect prediction). In a realistic scenario like ours—where we used synthetic data—you will probably get results that look worse than those during training (see Figure 6-12). That's totally fine and expected.

CHAPTER 6   AUDIO WAKE WORD DETECTION WITH EDGE IMPULSE

SAMPLE NAME	EXPECTED OUTC...	LENGTH	ACCURACY	RESULT
01J56ARQQ7...	wakeword	1s	100%	1 wakeword
01J56AQGJP...	wakeword	1s	100%	1 wakeword
01J56AS8Z4...	wakeword	1s	0%	1 noise
01J56ATX9PF...	wakeword	1s	100%	1 wakeword
01J56ASJ9NX...	wakeword	1s	100%	1 wakeword
01J56AV3SB...	wakeword	1s	100%	1 wakeword
01J56ATK5P...	wakeword	1s	0%	1 uncertain
01J56AV0HF...	wakeword	1s	100%	1 wakeword

Figure 6-12. Edge Impulse wake word detection test results

While most wake words are picked up correctly, there are some errors (the row with the *noise* label) and even a few *uncertain* classifications. These are classifications where the wake word is detected but with a confidence that is considered too low to be reliable. You can configure the "acceptable" confidence by clicking the gear icon to the right of the **Classify all** button and selecting **Set confidence thresholds**. I usually go for at least 0.6.

Take the time to manually inspect some errors (click the three dots to the right of the row and select **Show classification**) to understand why the model failed and if there's a recurring pattern in the samples misclassified. If you feel like the model does not match your expectations in terms of accuracy, go back to the training page and tune the parameters introduced in the previous section (model architecture, number of training cycles, learning rate). Repeat until you get better results.

> **Tip** If you still can't get good results no matter how hard you try to tune the training process, the reason may be a bad dataset. Take time to collect more data or move some of the samples that failed to classify to the training set (click the three dots to the right of the red rows and select **Move to the training set**), then train the model again.

Deployment

Download the model in the Arduino library format (see Chapter 4 for reference) and extract the ZIP file into your Arduino libraries folder. If you named your project keyword_spotting, the library is named keyword_spotting_inferencing.

Continuous Classification

STFT works on small windows of data (usually 20–40 ms), and its results are used to form the Mel spectrogram. When a new chunk of data arrives, the eldest one is discarded to make room for the latest. Recomputing the FT again for every chunk would waste time and resources since their result would not change.

Edge Impulse has a built-in mechanism to handle this *continuous* stream of data that you can leverage to achieve a near real-time classification speed. To fully exploit this technique, match EI_CLASSIFIER_SLICES_PER_MODEL_WINDOW in Listing 6-8 with the same value you set on the Impulse Design page.

Listing 6-8. Arduino Sketch for Keyword Spotting Using Edge Impulse

```
/**
 * Listing 6-8: Audio wake word detection using Edge Impulse
 *
 * Required hardware: Arduino Nano BLE Sense
 */
// next line *must* match with the window increase value in the
Impulse Design page from EI Studio
#define EI_CLASSIFIER_SLICES_PER_MODEL_WINDOW (1000 / 250)

#include <PDM.h>
// replace with the correct name of the library
// downloaded from Edge Impulse Studio
#include <keyword_spotting_inferencing.h>
#include <tinyml4all.h>

tinyml4all::PDMicrophone mic;
tinyml4all::Impulse impulse;

void setup() {
  Serial.begin(115200);
  while (!Serial);
  Serial.println("Keyword spotting with Edge Impulse");

  // match the volume with the one used for data collection!
  mic.frequency("16 khz");
  mic.volume(30);
  mic.begin();
  // init Edge Impulse library
  impulse.begin();
  // while testing, you can enable verbose output by
  // setting this value to true
```

CHAPTER 6 AUDIO WAKE WORD DETECTION WITH EDGE IMPULSE

```
  impulse.verbose(false);
}

void loop() {
  // await for audio data to be ready
  mic.await();

  // feed data to the impulse queue
  if (!impulse.queue(mic)) {
    Serial.println(impulse.error());
    return;
  }
  // skip non wake word
  if (impulse.label() != "wakeword")
    return;

  Serial.print("Wake word detected with confidence ");
  Serial.println(impulse.confidence());

  // customize here with your own logic
}
```

Compile the sketch (it takes a few minutes) and then upload it. Open the Serial Monitor and repeat the wake word aloud. You should see that the detection succeeded (see Figure 6-13).

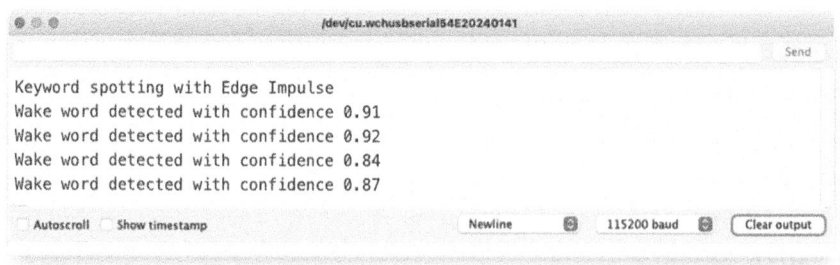

Figure 6-13. *Wake word detection output*

Inference time is not bad at all. It takes 77 ms to run a single inference. And since the PDM microphone captures data in a background task, you don't have waiting queues, nor do you risk missing chunks of data while performing inference.

Once this wake word detection project is working, you can collect more words to classify. Maybe you can implement a local music player assistant to detect Play, Pause, and Next to control your stereo using your voice—or Day and Night to control your home's lights. There are endless possibilities open to you.

> **Tip** Try to keep your wake words short (one or two seconds), or the processing time may be too long!

Summary

This chapter addressed a new type of data: audio. This is a very dense kind of data (many thousands of samples per second), but thanks to the tinyml4all library, we had no problems collecting training data fast. We were also able to leverage Generative AI to create a synthetic dataset in a matter of seconds.

The Edge Impulse low code platform made it easy to extract complex features using advanced algorithms (fast Fourier transform, Mel spectrogram, and MFCCs) and classify them using a convolutional neural network.

Deployment was pretty similar to Chapter 4. Reusing previous computations in the Mel spectrogram made it possible to achieve an astonishing inference speed of 13 classifications per second (77 ms per classification).

The next chapter addresses images using real-time object detection with an ESP32 camera.

CHAPTER 7

Object Detection with Edge Impulse

You have arrived at the last data type in this book: images. Images are so common in our everyday lives that we underestimate how complex and information-dense they are. Our brain has evolved for thousands of years to extract this information, even in bad environmental conditions (low light, occlusion, blurring, etc.). Training a computer program to replicate this same level of accuracy is a highly demanding task that still suffers compared with humans in many use cases.

Nevertheless, huge improvements have been achieved in this area, and today, we have many tools to perform object detection, which identifies objects of interest in images. This differs from pure classification: classifying an image means attributing a label to the *entire* image. Object detection, instead, is like classifying *regions* of the image, allowing the possibility of having more than one object of interest in the same image. Even many objects of different classes!

To lower the entry barrier to the tough problem of object detection on resource-constrained devices, we're once again leveraging the Edge Impulse platform. By now, you should have realized how versatile this platform is and how much it eases development. This chapter trains a model to recognize a single object of interest—a toy penguin (see Figure 7-1). Once familiar with this method, you can extend the project to recognize many different objects with little to no modifications.

CHAPTER 7   OBJECT DETECTION WITH EDGE IMPULSE

Figure 7-1. Object detection target

Hardware Requirements

You need a board with camera support. There are a few boards that come with a camera built-in.

- **Arduino-based**: Nicla Vision, Portenta H7 + vision shield
- **ESP32-based**: AiThinker, XIAO Sense, TTGO series, M5Stack series, ESP-EYE

Otherwise, you could try the more basic OV7670 camera that can work with many boards (e.g., Arduino Nano BLE Sense or Raspberry Pi Zero). But I discourage you from going this route because of the low quality and difficult wiring.

> **Caution** When looking for an ESP32 camera, beware that there are at least two chip variants. ESP32 is the "first" generation of chips with lower prices and specs. ESP32-S3 is the "next" generation: it costs a few more dollars, but the performance gain is huge. If possible, choose the S3 variant!

Software Requirements

The `tinyml4all` Arduino library has utility adapters for ESP32 and Arduino-based cameras, so the code stays the same. If you're going to use the OV7670 model, you need to write your own code to capture frames from it.

On the Python side, you only need the `tinyml4all` package (which you should have already installed).

Capture Data

To build the object detection dataset, you need to capture some images. How many images make a good dataset? As usual, the more, the better. But from my experience, you can achieve good results with as few as 50 images for each object.

> **Caution** You need to manually label each image you capture, so start low and add more later only if needed!

Many tutorials that you'll find on the web suggest that you use images from Google/the Internet to train your model. Some others suggest collecting data using your smartphone. Even if this approach could work well thanks to the generalization properties of the model, I strongly suggest you capture your own data with the same hardware that runs the inference (the microcontroller itself). It will better resemble your deployment scenario, and you will generally achieve higher accuracy.

> **Tip** Embedded camera sensors usually have poor quality and are impacted by environmental factors (e.g., poor illumination), unlike smartphones, which automatically correct most of these problems. A model trained on smartphone images may not perform well on embedded sensors' images.

Arduino Sketch to Collect Images

Let's send the captured frames over serial. Feel free to use an SD card if you prefer. (Most of the code won't change. You only need to write data on a file). Listing 7-1 has been tested on an ESP32-S3 camera: consult the book code repository if you're using a different model.

> **Tip** If using the ESP32 or ESP32S3, enable the external PSRAM option from the Tools menu of the Arduino IDE.

Listing 7-1. Send Images over Serial

```
/**
 * Listing 7-1: Collect images from camera
 *
 * Required hardware: ESP32 camera
 */
#include <tinyml4all.h>
```

```
tinyml4all::Camera camera;

void setup() {
  // increase Serial speed for fast data transmission
  Serial.begin(115200 * 2);
  while (!Serial);
  Serial.println("Collect images");

  // choose model from list
  camera.promptModel();
  // next times, you can set it directly by name
  // camera.setModel("aithinker");

  // configure camera for collection mode
  camera.collecting();

  // init camera (will throw an error on failure)
  camera.begin();
}

void loop() {
  // grab a new frame
  if (!camera.grab()) {
    Serial.println("Can't grab new frame");
    return;
  }
  // print frame to Serial (to be read from Python)
  camera.print();
  delay(1000);
}
```

CHAPTER 7   OBJECT DETECTION WITH EDGE IMPULSE

> **Tip**   The serial speed was increased for a faster transmission. With a one-second delay, the speed gain is not noticeable (it only saves a few milliseconds), but it comes in handy to collect images at a higher rate.

After the sketch is flashed, confirm that the board is streaming the image data over serial by opening the Serial Monitor. The output will look like garbage since it's binary data, not text. The important thing is that something gets printed. Now, it's time to run the Python collector script.

> **Tip**   If you get weird error messages or your board reboots when initializing the camera, double-check that you selected the correct model from the prompted menu.

Python Code to Read Images

On the Python side, the code looks similar to that used in previous chapters. As usual, you must configure the serial port, the number of samples to capture, and the destination folder (Listing 7-2). A preview of the camera is displayed, so you can fix the camera position if needed (see Figure 7-2).

Listing 7-2.   Collect Images over Serial

```
from tinyml4all.image import capture_serial

while True:
    # prompt user for object name and number of images
    object = input("Which object is this? ")
    num_samples = input("How many images to capture? ")
```

CHAPTER 7   OBJECT DETECTION WITH EDGE IMPULSE

```
    # exit if object name or duration is empty
    if not object or not num_samples:
        break
    capture_serial(
        # * is a wildcard character that matches anything
        # on Windows it will look like COM1 or similar
        port="/dev/cu.usb*",
        # must match with the Arduino sketch
        baudrate=115200 * 2,
        save_to=f"objects/{object}",
        num_samples=int(num_samples)
    )
Which object is this? penguin
How many images to capture? 30
Connected to serial port
100%|██████████████████████| 30/30 [00:33<00:00,  1.35s/it]
Disconnected from serial port
```

CHAPTER 7 OBJECT DETECTION WITH EDGE IMPULSE

Figure 7-2. Image collection preview window

Put the object in the correct position, possibly at the center of the frame and with good illumination, and complete the process. Next, open the destination folder and double-check that the images look good using your OS File Explorer.

You have to run at least two capture sessions.

- capture your object of interest
- capture background data (no object)

CHAPTER 7 OBJECT DETECTION WITH EDGE IMPULSE

The *background* data works the same as the *unknown* class introduced in Chapter 6; it hints at the model of what the object *doesn't* look like. Feel free to collect images of your surroundings, including chairs, computer monitors, walls, and windows. The more varied the images, the better.

Edge Impulse Data Acquisition

After you have collected a good amount of quality data, move on to https://edgeimpulse.com and create a new project named object-detection. As you saw in Chapter 6, you must upload your files to start (see Figure 7-3). The process is identical to our audio project.

1. Select either individual files or entire folder.
2. Enter the correct label manually.
3. Tick the **Automatically split between training and test testing** checkbox.

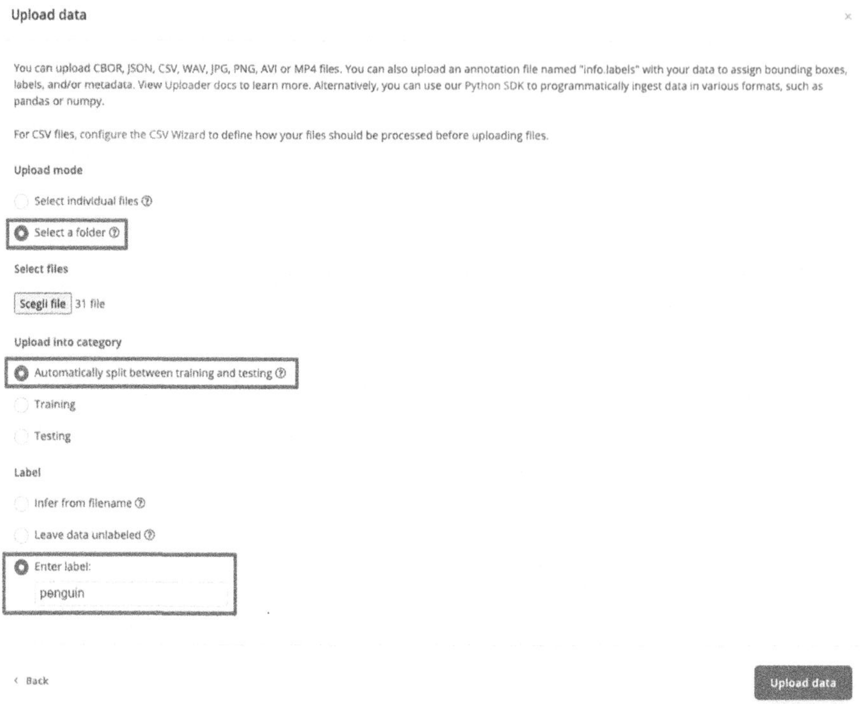

CHAPTER 7　OBJECT DETECTION WITH EDGE IMPULSE

Figure 7-3. Edge Impulse data upload form

But don't do this yet! Before you upload all the images you collected, read the next paragraph first!

You may be tempted to upload all your files while at the upload form. That would be a bad move. This is an object detection project, but you haven't yet defined which one our object of interest is inside Edge Impulse. **First, upload the images of your object of interest** (the penguin toy, in my case). After the upload is done, you are prompted if this is an object detection project (see Figure 7-4). Click **Yes**.

CHAPTER 7 OBJECT DETECTION WITH EDGE IMPULSE

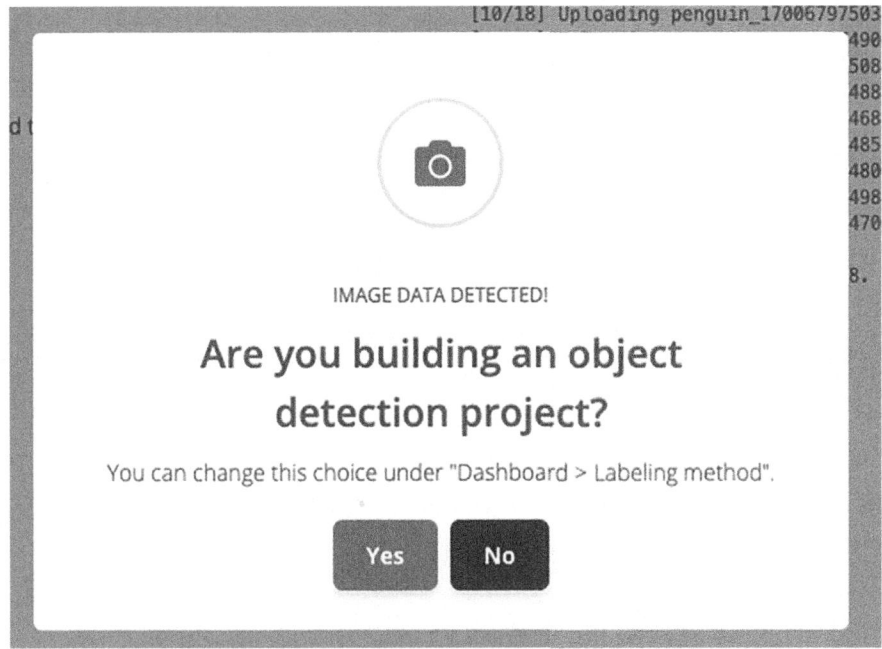

Figure 7-4. Edge Impulse labeling method selection

Instead of uploading more images, close the form and click **Labeling queue** from the top menu (see Figure 7-5).

Figure 7-5. Edge Impulse labeling queue menu item

A new window opens where you can see your images in the center and the message: "Use your mouse to drag a box around an object to add a label. Then click **Save labels** to advance to the next item." That's what you will do for all the images: draw a box around each object of interest and assign it the correct label. If you leave the **Track objects between frames** selected, the process will be faster because, typically, you only need to do minor fixes to the bounding box that is automatically positioned (see Figure 7-6).

CHAPTER 7   OBJECT DETECTION WITH EDGE IMPULSE

Figure 7-6. Edge Impulse bounding box drawing tool

Tip   If you upload all the images at once, the labeling queue mixes the different objects, and you will lose a lot of time drawing the bounding boxes!

If you have more objects, upload them one at a time and do the labelling for each. After you finish the objects of interest images, upload the *background* images. Then open the labeling queue again and click **Save labels** without drawing anything. Since this is background, there's nothing to highlight.

Feature Engineering

Feature engineering for image classification and object detection is inexistent. Or better, it is built into the model. As stated in Chapter 6, when it comes to images, the reference neural network architecture is *convolutional*. Even if it is pretty hard to "see" what happens exactly in the middle layers of a convolutional neural network, an intuitive qualitative interpretation is that those filters (each layer has many filters) learn to

CHAPTER 7   OBJECT DETECTION WITH EDGE IMPULSE

recognize patterns and features more and more complex as the input moves forward the network. The first layers often learn basic features like edges and corners, while the deeper layers learn to recognize basic shapes and textures (see Figure 7-7).

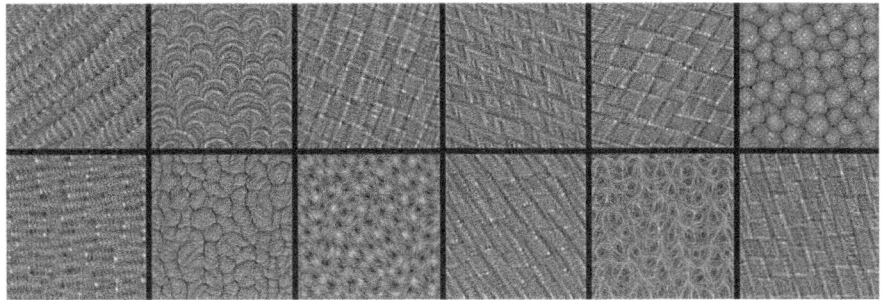

Figure 7-7. Activation maps of inner layers of a 2D convolutional neural network

The whole point of a convolutional neural network is that its modeling power is so efficient that it figures out how to best detect recurring patterns in the different classes. What a couple of decades ago was a manual task has been completely automated today.

Impulse Design

There's not much you can do here because the processing and classification blocks are fixed for this type of data (see Figure 7-8). But you still must make two important decisions.

- **Image size**: Your camera can capture frames at different resolutions. A pretty common resolution is 320×240 pixels. Despite this resolution looking pretty small compared to our everyday experience of photos (several megapixels), it is in the high range of the image analysis spectrum. Most "mobile" neural networks (meant to

CHAPTER 7   OBJECT DETECTION WITH EDGE IMPULSE

run on low-end CPUs like smartphones from a few years ago or a Raspberry Pi) work on 192×192 images or lower. Your board can't even handle that size probably!

To avoid out-of-memory errors, I suggest starting with a size of 80×80 in RGB mode or 96×96 in grayscale mode. You can always come back later and increase the resolution if the model performs badly and you have enough resources for a larger model. I discourage you from trying anything higher than 128×128.

- **Color depth** (on the Image page): Most cameras can capture images in RGB or grayscale modes. Even if your captured frames are in RGB, you can force the model to process them in grayscale. Why would you do that? For memory constraints reasons, of course. Intuitively, you can expect an RGB model to require three times the memory of a grayscale model to store its intermediate results and triple the time to execute. Unless your board has plenty of RAM, this can be prohibitive and prevent your model from working.

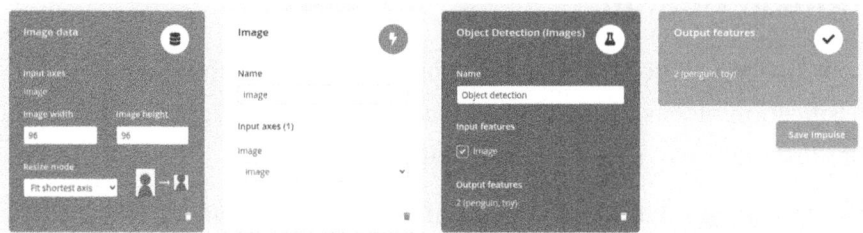

***Figure 7-8.** Edge Impulse Design page for object detection*

When it comes to the model configuration step, Edge Impulse provides two alternative architectures that fit inside a microcontroller. (They offer more options for Raspberry Pi/Jetson Nano/PC targets).

- FOMO MobileNetV2 **0.1**

- FOMO MobileNetV2 **0.35**

Note FOMO stands for "Faster Objects, More Objects" and is the branded name that the Edge Impulse team gave to their custom, optimized implementation of object detection.

First, both models are derived from MobileNetV2[1], a lightweight and efficient convolutional neural network architecture designed for mobile devices (like our smartphones). However, smartphones are still orders of magnitude more powerful than a microcontroller (gigabytes of RAM, CPU frequencies in the gigahertz range), so that network is still too heavyweight to be considered "tiny," as defined in this book.

That's where the decimal number after its name comes into play. It is called a width multiplier and controls the number of filters in the network. As briefly introduced in Chapter 6, a convolution layer is made of many filters (or kernels). The width multiplier decimates the number of these filters starting from the original count. A width multiplier of 0.1 means that only 10% of the original number of filters is used; 0.35 implies roughly one-third of the count.

The huge achievement is that the reduction in the number of total weights of the network is *quadratically* proportional to this multiplier. Since the number of weights in a convolutional layer depends on the number of input channels, the number of output channels, and the kernel size, scaling the number of filters by α (0.1 or 0.35) reduces the number of weights quadratically for the following reasons.

- The number of output channels is reduced by α.

- The number of input channels to subsequent layers is also reduced by α.

- The kernel size is constant.

Thus, the 0.35 variant only holds about 12% of the original MobileNetV2 weights, while the 0.1 variant reduces the size to *only 1% of the original!* Of course, you can expect the larger version to show better accuracy.

Which one you choose depends on a couple of factors.

- **Memory constraints**: Low-resources boards cannot handle the larger version, forcing your choice. If you're using RGB images, keep in mind that the problem only gets worse.

- **Time constraints**: If you're building a near real-time project (e.g., an autonomous robot) and need the highest responsiveness possible, you're more interested in running fast than accurately. On the other hand, if you can afford to spend up to 500 ms to analyze a single frame and have enough memory, the 0.35 version is more reliable.

For my project, I chose the 0.1 variant. If your dataset has only high-quality images, you can expect a high accuracy even from this smaller model. You can choose a different model on the Object Detection page (menu on the left) by clicking **Choose a different model**, as shown in Figure 7-9.

CHAPTER 7   OBJECT DETECTION WITH EDGE IMPULSE

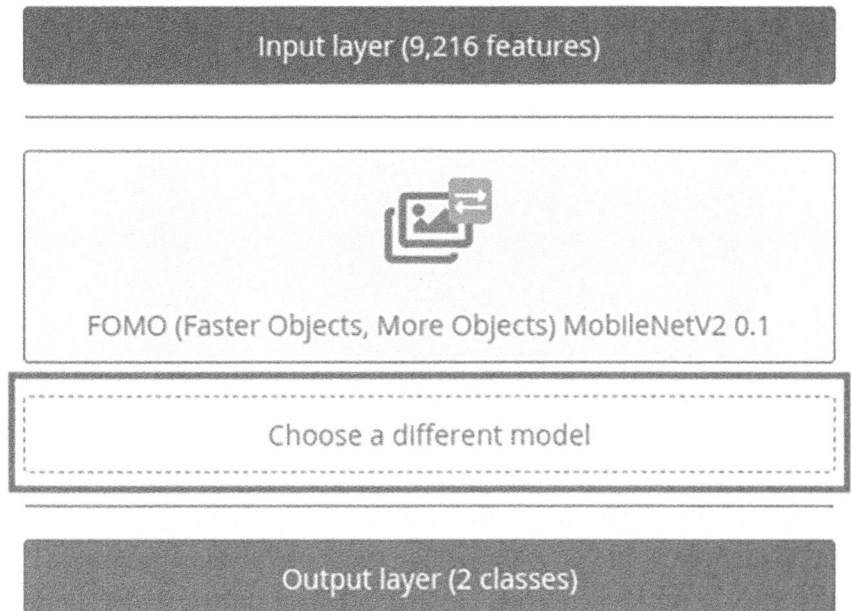

Figure 7-9. Choose FOMO model

Run the training and then move to the model testing page.

Testing

If the confusion matrix on the training page satisfies your expectations, move to the Model Testing page (from the menu on the left) to assess the model accuracy on images it has never seen during training or validation. Click the **Classify all** button and wait for the results. In the best scenario, all the rows are green (indicating a correct prediction; see Figure 7-10).

CHAPTER 7   OBJECT DETECTION WITH EDGE IMPULSE

SAMPLE NAME	EXPECTED OUTCOME	F1 SCORE
penguin_1700...	penguin	100%
penguin_1700...	penguin	100%
penguin_1700...	penguin	100%

Figure 7-10. Edge Impulse model results on test set

On more complex datasets, it is more likely that some objects will be classified incorrectly—or not detected at all! You should pay a lot of attention to these results (way more than those on the training page) because they're the most reliable indication of how the model will perform after being deployed on new images it has never seen.

Take the time to manually inspect some errors (click the three dots to the right of the row and select **Show classification**). If you feel like the model does not match your expectations in terms of accuracy, go back to the training page and tune the parameters introduced in the previous chapter (model architecture, number of training cycles, learning rate). Repeat until you get better results.

Caution   If you don't get good results no matter how hard you try to tune the training process, the reason may be a bad dataset. Collect more images and double-check that the bounding boxes you draw are correct.

CHAPTER 7   OBJECT DETECTION WITH EDGE IMPULSE

Deployment

To deploy the object detection model inside an Arduino sketch, download the model in the Arduino library format (see Chapter 4 if you cannot find how to do this) and extract the zip into your Arduino libraries folder. If you named your project object-detection, the library is named object-detection_inferencing.

Create a new Arduino project and paste the code in Listing 7-3.

Listing 7-3. Arduino Sketch for Object Detection Using Edge Impulse

```
/**
 * Listing 7-3: Object detection using Edge Impulse.
 *
 * Required hardware: ESP32 camera
 *
 * Notes: when using ESP32 boards, don't forget
 * to enable PSRAM from the Tools menu!
 */
#include <object-detection_inferencing.h>
#include <tinyml4all.h>

tinyml4all::Camera camera;
tinyml4all::Impulse impulse;

void setup() {
  Serial.begin(115200);
  while (!Serial);
  Serial.println("Run Edge Impulse Object Detection");

  // choose model from list
  camera.promptModel();
```

CHAPTER 7   OBJECT DETECTION WITH EDGE IMPULSE

```
  // next times, you can set it directly by name
  // camera.setModel("wroom-s3");

  // configure camera for inference mode
  camera.inferencing();
  // init camera
  camera.begin();

  // init Edge Impulse model
  impulse.begin();
}

void loop() {
  // gran new frame
  if (!camera.grab()) {
    Serial.println("Can't grab new frame");
    return;
  }

  // run impulse
  if (!impulse.run(camera)) {
    Serial.println(impulse.error());
    return;
  }

  // if no object is detected, return
  if (impulse.count() == 0)
    return;

  // print how many objects have been detected
  Serial.print("Found ");
  Serial.print(impulse.count());
  Serial.println(" object(s)");
  // print objects with coordinates
```

CHAPTER 7   OBJECT DETECTION WITH EDGE IMPULSE

```
for (int i = 0; i < impulse.count(); i++) {
  auto object = impulse.object(i);

  Serial.print(" > ");
  Serial.print(object.label);
  Serial.print(" at coordinates ");
  // cx, cy are the coordinates of the
  // center of the object
  Serial.print(object.cx);
  Serial.print(", ");
  Serial.print(object.cy);
  Serial.print(" (confidence ");
  Serial.print(object.confidence);
  Serial.println(")");
  }
}
```

Compile the sketch—it takes a few minutes the first time—and upload. Open the Serial Monitor and put your object of interest in front of the camera. You should see that the detection succeeded when you put the object of interest in front of the camera (see Figure 7-11).

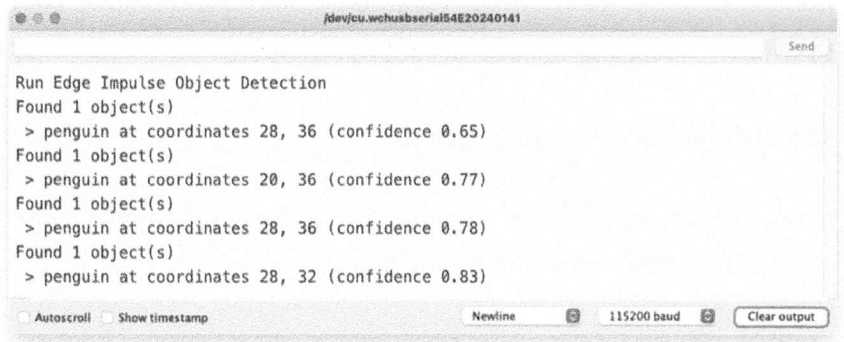

Figure 7-11. Inference serial output

CHAPTER 7   OBJECT DETECTION WITH EDGE IMPULSE

To give you some figures on the execution time of this project, consider that an ESP32 (non-S3) takes ~280ms to run a MobileNetV2 0.1 model on 64×64 RGB frames. An ESP32 S3 takes ~35ms. The Arduino Nicla Vision is even faster at 20ms!

The accuracy is far from 100%, but that was expected given the poor sensor quality and the low model resolution. When implementing production firmware, you must consider that the model may miss some frames or detect objects that aren't really there.

Visual Debugging

Getting the detection results printed on the Serial Monitor is the quickest way to assert that the model is working. But is it also possible to *visually* inspect the results? Can you get a nice preview of what the camera sees and where the object of interest is located? Yes, of course.

The following Arduino sketch combines Listing 7-1 and Listing 7-3. Let's print the frame *and* the prediction results over the serial connection (see Listing 7-4) to be later analyzed and displayed from a Python script.

Listing 7-4. Arduino Sketch to Debug Object Detection Results

```
/**
 * Listing 7-4: Debug object detection results
 *
 * Required hardware: ESP32 camera
 *
 * Notes: when using ESP32 boards, don't forget
 * to enable PSRAM from the Tools menu!
 */
#include <object_detection_inferencing.h>
#include <tinyml4all.h>
```

CHAPTER 7   OBJECT DETECTION WITH EDGE IMPULSE

```
tinyml4all::Camera camera;
tinyml4all::Impulse impulse;

void setup() {
  // increase Serial speed for faster image transfers
  Serial.begin(115200 * 2);
  while (!Serial);
  Serial.println("Debug object detection results");

  // set it directly by name
  camera.setModel("wroom-s3");

  // configure camera for inference mode
  camera.inferencing();

  // init camera
  camera.begin();

  // init Edge Impulse model
  impulse.begin();
}

void loop() {
  // gran new frame
  if (!camera.grab()) {
    Serial.println("Can't grab new frame");
    return;
  }

  // print frame
  camera.print();
  // run impulse
```

CHAPTER 7 OBJECT DETECTION WITH EDGE IMPULSE

```
  if (!impulse.run(camera)) {
    Serial.println(impulse.error.msg);
    return;
  }

  if (impulse.count() == 0)
    return;

  Serial.print("Found ");
  Serial.print(impulse.count());
  Serial.println(" object(s)");
  // print objects with coordinates
  for (int i = 0; i < impulse.count(); i++) {
    auto object = impulse.at(i);

    Serial.print(" > ");
    Serial.print(object.label);
    Serial.print(" at coordinates ");
    // cx, cy are the coordinates of the
    // center of the object
    Serial.print(object.cx);
    Serial.print(", ");
    Serial.print(object.cy);
    Serial.print(" (confidence ");
    Serial.print(object.confidence);
    Serial.println(")");
  }
}
```

CHAPTER 7 OBJECT DETECTION WITH EDGE IMPULSE

Next, create a new Python script (see Listing 7-5). The script is similar to Listing 7-2, but this time, you won't be prompted for the name of the object since it is for visualization purposes only.

Listing 7-5. Display Object Detection Results

```
from tinyml4all.image import debug_serial
debug_serial(
    # * is a wildcard character that matches anything
    port="/dev/cu.usb*",
    # must match with the Arduino sketch
    baudrate=115200 * 2
)
```

Run the script, and a new window should appear with a preview of the camera stream (see Figure 7-12) and a red dot on the object of interest (if any).

CHAPTER 7 OBJECT DETECTION WITH EDGE IMPULSE

Figure 7-12. Object detection results debug window

Congratulations! You completed your first object detection project and the last hands-on from this book. Take some time to celebrate your achievements.

Summary

This last "beginner-friendly" chapter leveraged the low-code Edge Impulse platform to implement state-of-the-art real-time object detection on our camera-equipped board. The workflow was linear and intuitive: capture

CHAPTER 7   OBJECT DETECTION WITH EDGE IMPULSE

images of our object of interest, label them using a simple GUI by drawing bounding boxes around the object, train a neural network to do the heavy work, and deploy the network back to our microcontroller.

Every operation on the board was just a couple of lines away, thanks to the `tinyml4all` library, so you have total control over your firmware. You can integrate custom logic (e.g., activate a relay, turn on an LED) when objects are detected, and we even have access to the (rough) position of the object for more advanced applications (e.g., pan/tilt servo motor, autonomous driving robot).

The final chapter takes a huge step forward and digs into the ins and outs of artificial neural networks: how they work, how to create and train one from scratch, and finally, how to deploy it back to our Arduino microcontroller without relying on external platforms.

CHAPTER 8

TensorFlow from Scratch

So far, we've either implemented "traditional" machine learning with manual feature engineering and classifiers/regressors models or leveraged the low-code Edge Impulse platform to unleash the power of neural networks without digging into the nitty gritty details of code. But this book aims to turn you into a proficient TinyML practitioner, and it won't be complete without a hands-on chapter on how to work with neural networks by writing code.

Several Python frameworks implement neural networks, and one of the most used—and the only one that integrates well with the Arduino environment—is TensorFlow [1]. TensorFlow is a project managed by Google that powers many products that run state-of-the-art deep learning algorithms. This chapter focuses on four architectures that can be ported to our embedded hardware: multilayer perceptron for tabular data classification, recurrent neural networks and 1D convolutional neural networks for time series classification, and 2D convolutional neural networks for image classification.

Let's leverage the project data already collected (or use toy datasets or synthetic data) and focus on the definition, training, and deployment part of neural networks using Python and the `tinyml4all` package.

CHAPTER 8　TENSORFLOW FROM SCRATCH

Required Hardware

This chapter focuses on teaching you how to train and deploy neural network architectures.

For this reason, the only hardware requirement is a board capable of running TensorFlow models. Due to compatibility easiness, the Arduino code in this chapter runs equally well on the following.

- ARM Cortex-M boards (Arduino Nano and Nicla families, Raspberry Pi Zero)
- ESP32 (both S3 and non-S3 variants)

If you have a different chipset, refer to the specific documentation from the board supplier.

Caution　TensorFlow models need a lot of RAM to work, so choose a board with 256 KB at minimum (even better 512 KB).

Required Software

On the Python side, apart from the tinyml4all library, you need the TensorFlow library. This is not included by default because it is large (>100 MB), and not everyone may find it useful.

To install it, open your terminal, activate your virtual environment, and type the following commands.

```
(tinyml) $ python -m pip install "tinyml4all[tensorflow]"
```

This installs a specific version of TensorFlow (2.15.1) because newer releases won't work with all the models described in this chapter. If you already have a later version installed, you must downgrade.

CHAPTER 8 TENSORFLOW FROM SCRATCH

Tip If you don't want (or can't) install TensorFlow on your PC, you can run the following Python scripts in Google Colab (https://colab.research.google.com), a cloud computing environment from Google. The free tier even provides a GPU, giving you a lot of power.

Run the following line in the first cell to set up all the dependencies.

```
! python -m pip install "tinyml4all[tensorflow]"
```

On the Arduino side, you must install an additional library. Again, TensorFlow is a large dependency that may slow the compilation down when not needed, so it's not included in the tinyml4all Arduino library.

Open the Library Manager from the Arduino IDE, search for tensorflow-runtime-universal, and install it (see Figure 8-1).

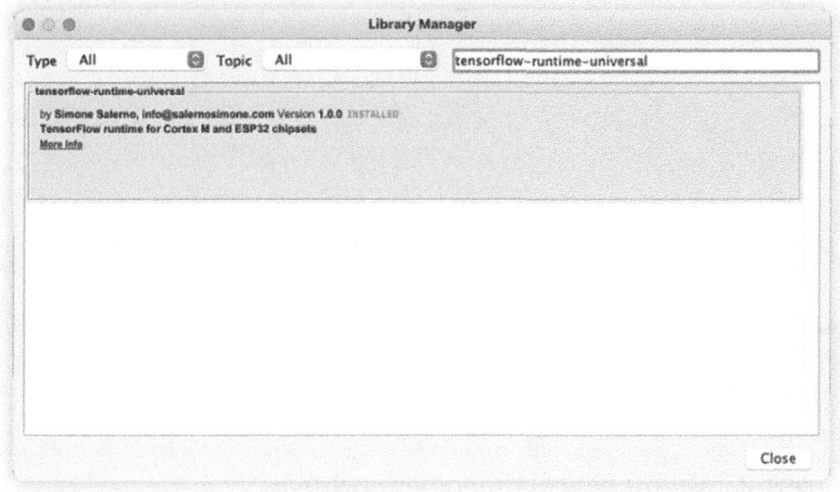

Figure 8-1. Install tensorflow-runtime-universal Arduino library

257

CHAPTER 8 TENSORFLOW FROM SCRATCH

Neural Network Structure

Neural networks come in many shapes. Since this book is not about deep learning, it only focuses on the simplest architecture, called *sequential*, where a network is defined as a sequence of *layers*. Layers are, in simple terms, groups of *neurons*.

The neuron is the fundamental atomic component of a network. There are many types of neurons, each encapsulating a different computation logic. By leveraging the ensemble of (possibly different types of) neurons, the entire network can model complex input/output relations (see Figure 8-2).

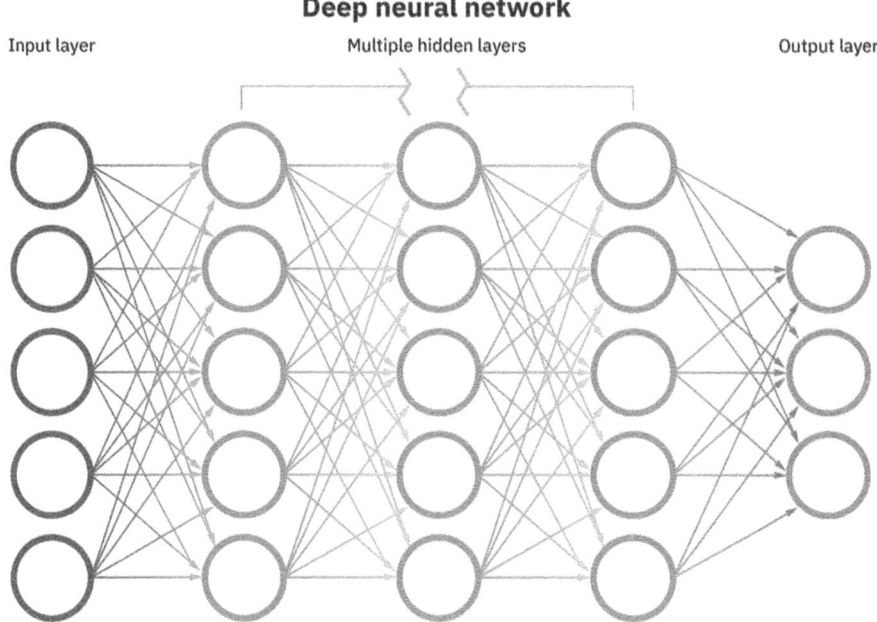

Figure 8-2. Neural network topology example

Every network architecture is composed of three essential parts, which are always present.

- **Input layer**: The very first layer of a network is represented by its inputs. These values come from the *outside world* and are not really under the control of the network. Input values can be tabular data, audio, time series, images, videos, and so on. What characterizes the input is its *shape* (how many dimensions it has).

- **Output layer**: This is the last layer of the network and contains the results we're interested in. For classification tasks, this layer usually contains the probability that the input sample belongs to each possible class. For regression tasks, this layer contains the predicted values. For this reason, the output layer shape is also somehow out of our control and not arbitrarily configurable.

- **Hidden layers**: These are all the layers that are placed between the input and output ones. Depending on the network topology, they can contain a few or many layers (even tens or hundreds), all the same type or from different types, with varying sizes and complexity. **This is the part you have control over.** The role of this part is to allow the network to learn the relation between the inputs and the outputs. More layers lead to more complex relationships at the cost of more memory/computation requirements.

Caution When working with TinyML, you want to keep the hidden layers as lean as possible to respect the hardware constraints.

Forward Pass

Given an input sample, going through the network layers to produce a result is called *inference* or *forward pass* (it helps imagine the network developing horizontally from left to right). The data flow happens sequentially: the first hidden layer processes the input values. Its outputs become the input for the second hidden layer, and so on, until we reach the end at the output layer (*non-sequential* architectures may have different execution paths).

Backward Pass

Now that you know how to produce an output, you need a way to train the network to produce *accurate* outputs. This is achieved with a *backward pass* (also called *back-propagation*). The intuitive logic of back-propagation is the following.

1. Each neuron in the network is initialized with semi-random values, so they hold no knowledge about the data. They're in a blank state.

2. Given an input, you compute its corresponding output. Presumably, this output will be wrong since the network has never seen this data.

3. You then compute the error (e.g., the current output minus the expected output for regression) and go back to each layer, saying, "At your current state, you produced this error. Try to update your state so that the error is reduced." This update process varies for each neuron type.

4. By repeating this process. for each input sample, things should improve. However, only showing each input a single time may not be enough for the network to work fine (especially if there are many neurons whose interactions can create waterfall effects). For this reason, this process is repeated many times (called *epochs*). Usually, the larger the network, the more epochs are required to achieve good results.

This is a very broad description of how neural networks work. Each specific type of neuron has a different forward and backward (update) logic. Let's explore the most common neurons. and architectures.

Multilayer Perceptron

As the name suggests, a multilayer perceptron (MLP)—also called a *fully connected network*— is made of many layers of *perceptrons*. A perceptron is a (simplified) mathematical representation inspired by the human brain cells (see Figure 8-3).

CHAPTER 8 TENSORFLOW FROM SCRATCH

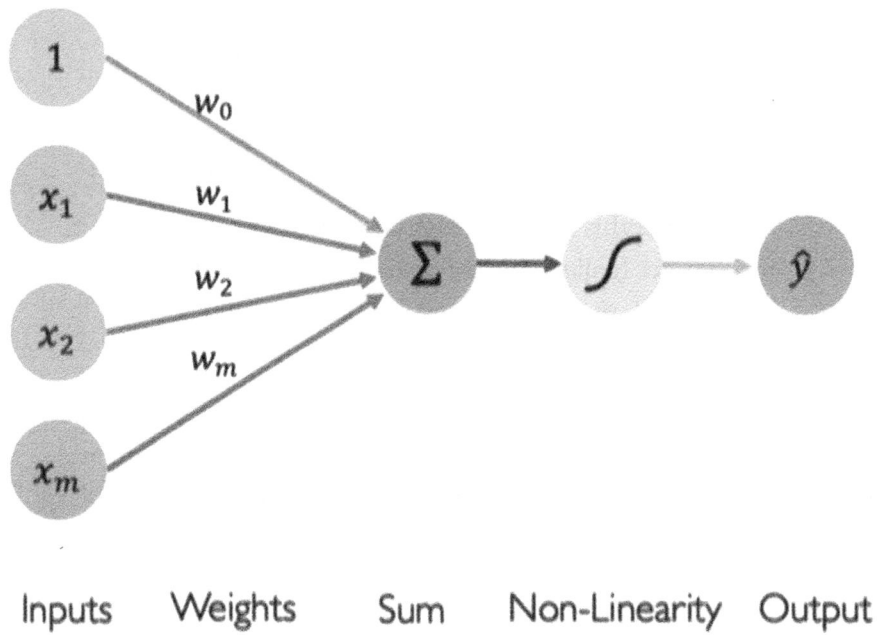

Figure 8-3. *Diagram of a perceptron neuron*

In simple terms, the brain cells (a.k.a. *neurons*) receive signals from other neurons with connections of varying strengths (*synaptics*) and process them. If the result exceeds a given threshold, they produce an output signal (*action potential*). In 1957, the psychologist Frank Rosenblatt [2] had the intuition to replicate this mechanism, in a simplified form, into an electronic device. Today, the foundational principle of a perceptron remains unchanged.

- The perceptron receives *n* inputs, from the *outer world* (e.g., sensors) or other neurons.
- Each input has a *weight* associated. The processing inside the perceptron consists of the dot product between the weights and the inputs, plus an offset (called *bias*).
- To simulate the action potential firing, the result of the dot product undergoes an *activation function*, which introduces a non-linearity in the model.

The most important element of the perceptron is probably the activation function. Without this, the entire model would be linear despite the number of neurons. With the introduction of a non-linearity at each neuron, the model is able to learn more complex relations. The most common activation function is a *rectified linear unit* (ReLU), which truncates all negative values to 0, as explained in the following formula.

$$ReLU(x) = max(0, x)$$

The *multilayer* part of an MLP is achieved by stacking groups of perceptrons one after the other in a very schematic topology where each neuron in a layer is connected to each other in the next layer (thus fully connected, as seen in Figure 8-4).

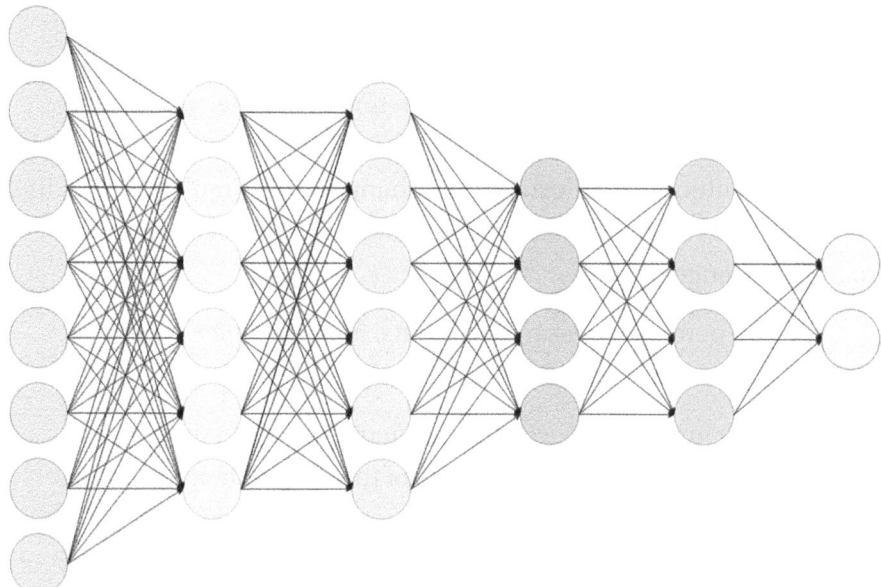

Figure 8-4. *Fully connected topology*

Now, let's make a quick example to investigate how the number of layers and neurons affects the total number of weights. Assume we have a network with ten inputs, three outputs, and three hidden layers with 8, 16, and 24 neurons, respectively. You can easily compute the total number of weights that comprise this network.

- The first hidden layers receive 10 inputs, so each perceptron has 10 weights + 1 bias. With 8 perceptrons, there are 8 * 11 = 88 weights.

- The second hidden layers receive 8 inputs, so 16 * (8 + 1) = 144 weights.

- The third hidden layer receives 16 inputs, so 24 * (16 + 1) = 408 weights.

- The output layer receives 24 inputs and has three neurons, so 3 * (24 + 1) = 75 weights.

CHAPTER 8 TENSORFLOW FROM SCRATCH

Total weights = 88 + 144 + 408 + 75 = 715.

As you can see, the number of weights can quickly grow and add up as the number of layers and neurons inside each layer grows. You can easily find networks with hidden layers of 64 or 128 neurons, which produce thousands of weights each. This may not be a problem when running on your desktop PC, but it *can easily saturate the available RAM* on your microcontroller. Always remember this when trying out a fully connected topology for your project!

As a companion drawback, it is also evident that *the more weights*, the more computations the CPU must perform, so *the longer it takes* to produce a result on your slow embedded hardware.

How to Train a Multilayer Perceptron

Now that you know what makes an MLP, it's time to create and train one in Python. Since fully connected networks work fine with tabular data, let's use a toy dataset called Iris [3], which classifies the species of an Iris flower by its sepal and petal width and length (see Figure 8-5).

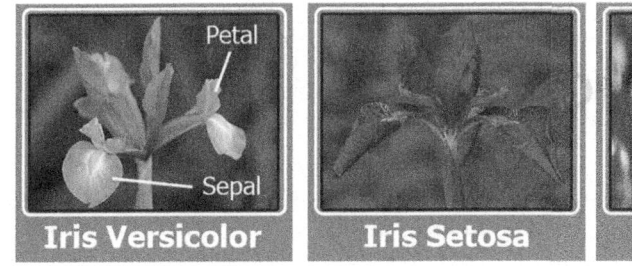

Figure 8-5. Iris dataset samples

This is a toy dataset that is very easy to classify. Using a random forest classifier, you can expect 95% or higher accuracy with no pre-processing. Listing 8-1 introduces the MLP class and explains how to add layers, train the network, and get a classification report. Figure 8-6 displays the loss and accuracy of the network.

CHAPTER 8   TENSORFLOW FROM SCRATCH

Listing 8-1. Train an MLP on the Iris Toy Dataset

```
from tinyml4all.tensorflow import MLP
from tinyml4all.tensorflow.layers import Perceptron
from tinyml4all.datasets import Iris

# Iris is an instance of tinyml4all.tabular.
classification.Table

print(Iris)

# instantiate a new network for the Iris dataset
mlp = MLP()
# 8, 16, 24 are the number of perceptrons for each layer
mlp.add(Perceptron(8))
mlp.add(Perceptron(16))
mlp.add(Perceptron(24))

# display network architecture
# X is the input data
# Y are the labels
# task can either be "classification" or "regression"
print(mlp.compile(X=Iris.numeric, Y=Iris.targets.values,
task="classification"))
Model: "sequential"
```

Layer (type)	Output Shape	Param #
dense (Dense)	(None, 8)	40
dense_1 (Dense)	(None, 16)	144
dense_2 (Dense)	(None, 24)	408
dense_3 (Dense)	(None, 3)	75

Total params: 667 (2.61 KB)
Trainable params: 667 (2.61 KB)
Non-trainable params: 0 (0.00 Byte)

```
# train neural network and display accuracy plot
mlp.fit(X=Iris.numeric, Y=Iris.targets.values, epochs=50,
plot=True)

# print accuracy on the training/validation set
mlp.classification_report()
```

```
              precision    recall  f1-score   support

      setosa       1.00      1.00      1.00        50
  versicolor       1.00      0.92      0.96        50
   virginica       0.93      1.00      0.96        50

    accuracy                           0.97       150
   macro avg       0.98      0.97      0.97       150
weighted avg       0.98      0.97      0.97       150
+--------------------+--------+------------+-----------+
| True vs Predicted  | setosa | versicolor | virginica |
+--------------------+--------+------------+-----------+
|             setosa |   50   |      0     |     0     |
|         versicolor |    0   |     46     |     4     |
|          virginica |    0   |      0     |    50     |
+--------------------+--------+------------+-----------+
```

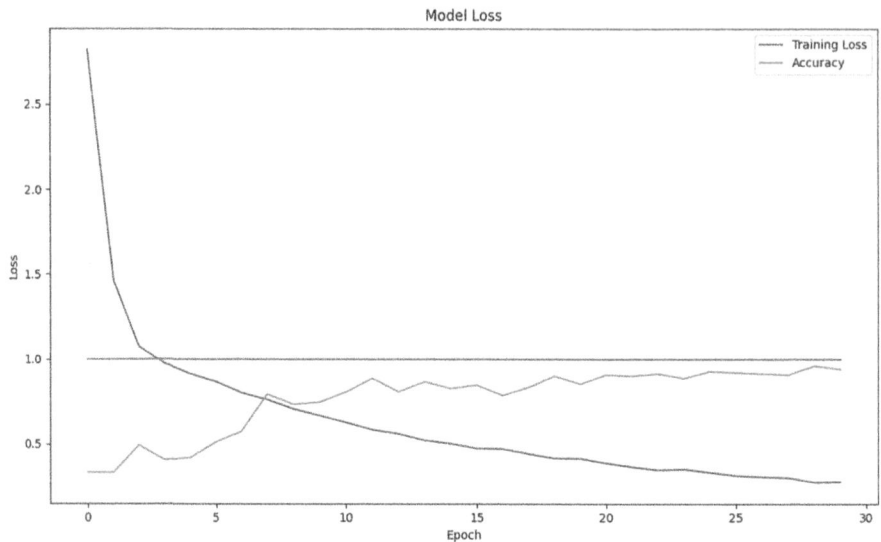

Figure 8-6. *MLP loss and accuracy plot*

Tip If you have used TensorFlow before, you may find the class names in the code disappointing. They were chosen to be consistent with the text in this book, but they're just aliases for built-in layers. Perceptron maps one-to-one to dense and MLP to sequential.

I want you to notice one thing: we added three perceptron layers (with 8, 16, and 24 neurons each), but the network dump is showing one additional dense layer with three neurons. Why is that? All the architectures showcased in this chapter (MLP, RNN, CNN1D, and CNN2D) perform some manipulations to the layers before building the final model. In this case, the MLP class injected one final layer with as many neurons as the number of classes (three for *setosa, virginica,* and *versicolor*) with a *softmax* activation function [4] that produces class membership probabilities.

The output of the network is not a single value as you might expect, indicating the predicted class, but a vector of values (probabilities) ranging from 0 to 1, one value per class. If you need a single output, you can pick the class with the highest probability. This scheme allows for a more granular output handling because you can introduce an *uncertain* metaclass if the confidence is below a given threshold (like Edge Impulse does in its Model Testing page).

> **Tip** If you used TensorFlow before, you might have noticed that you don't have to define the loss function nor the metrics: this is all handled for you with `task="classification"`. If you want to take total control over the `compile()` parameters, you can still supply them and they take precedence over the defaults.

Built-in Validation

To train a neural network effectively, you need a validation set. For this reason, when you call `fit()`, a split between train and validation sets is automatically created. The default is 80% for training and 20% for validation.

You can change the validation ratio manually if you'd like.

```
# use 30% of data as validation
```

```
mlp.fit(X=Iris.numeric, Y=Iris.targets.values, validation_data=0.3, epochs=50, plot=True)
```

Even if discouraged, you can disable validation altogether.

```
mlp.fit(X=Iris.numeric, Y=Iris.targets.values, validation_data=0, epochs=50, plot=True)
```

A test split is not automatically created. You can force the generation of a test split with the test_data argument.

```
# use 20% of data as test
mlp.fit(X=Iris.numeric, Y=Iris.targets.values, test_data=0.2,
epochs=50, plot=True)
```

Or, if you have an externally defined test dataset (e.g., loaded from another CSV file), you can use it instead.

```
# set external test dataset
test_dataset = Table.read_csv("test_dataset.csv")
X_train = Iris.numeric
Y_train = Iris.targets.values
X_test = test_dataset.numeric
Y_test = test_dataset.targets.values
mlp.fit(X_train, Y_train, test_data=(X_test, Y_test),
epochs=50, plot=True)
```

Note If you define a test while fitting, metrics and plots use it instead of the validation one.

How to Deploy a Multilayer Perceptron

To convert the MLP into Arduino-compatible C++ code, simply run the following.

```
mlp.convert_to("c++", save_to="IrisMLP.h")
```

The generated code encapsulates all the complex parts required to run TensorFlow on a microcontroller and exposes a deadly simple API (see Listing 8-2). The output of the example is shown in Figure 8-7.

Listing 8-2. Run MLP on Arduino

```
#include "./IrisMLP.h"

// ARENA is the amount of memory to reserve for the model.
// larger models need more memory, but there's not a
// formula to calculate the optimal value,
// is a trial-and-error process
#define ARENA 20000

tinyml4all::MLP<ARENA> mlp;
float setosa[] = {5.1, 3.5, 1.4, 0.2};
float virginica[] = {7.6, 3.0, 6.6, 2.1};
float versicolor[] = {6.8, 2.8, 4.8, 1.4};

void setup() {
    Serial.begin(115200);
    while (!Serial);
    Serial.println("TensorFlow MLP demo");

    // init network
    mlp.begin();
}
void loop() {
    // run network on given sample
    if (!mlp.predict(setosa)) {
        Serial.println(mlp.error());
        return;
    }
    // mlp.label holds the name of the predicted class
    // mlp.idx holds the numeric id of the predicted class
    // mlp .value holds the numeric output (for regression
       tasks only)
```

```
// mlp.confidence is the probability of the prediction
//                  (from 0 to 1)
// mlp .confidences  is an array with the confidences of
   all the classes
//              (one for each class for classification)
// mlp.outputsAsString returns each class' score, from
//              0 to 1 (classification only)
Serial.print("Predicted class ");
Serial.print(mlp.label);
Serial.print(" with confidence ");
Serial.println(mlp.confidence);

Serial.print(" > Scores: ");
Serial.println(mlp.outputsAsString());
delay(1000);
}
```

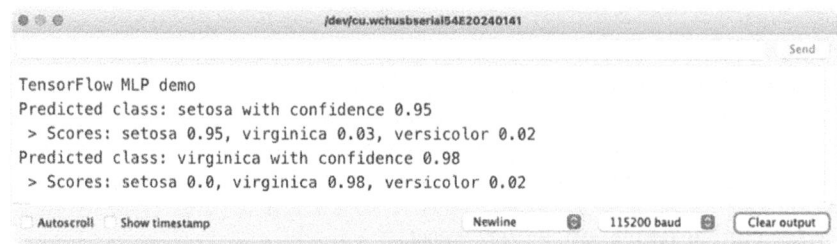

Figure 8-7. MLP output example

Note The perceptron is a ubiquitous operator. It appears as a glue between specialized operators (LSTM, convolution) and the output layer in almost every network topology, often in the final layers. You find it in every other type of network discussed in this chapter.

Deep Learning

Chapters 4 and 5 approached the classification of time series as a two-step process: (1) extract meaningful features from time series (either using the Fourier transform or time-domain descriptors) and (2) classify those features using a *tabular* classifier (fully connected network on Edge Impulse, random forest in Python).

With the rise in popularity and accessibility of deep learning techniques in recent years, a different approach has become widespread: embed the feature extraction inside the same neural network that does the classification. In this sense, there's an important paradigm shift. Before, we manually defined what *we thought* could make a good feature vector. Now, we delegate to a machine learning/deep learning model to construct the best features, driven by a training process on the input dataset. Depending on the dataset and the type of network, this approach can yield better accuracy than our manual work.

Deep Learning Disadvantages in the TinyML Context

Of course, there's a coin-flip: to be so expressive, the network has to grow in size and complexity. The fully connected network trained in Chapter 4 consisted of just two layers with a few neurons each (since the hard work was concentrated in the spectral analysis step). This simple topology was meant to work on heavily processed input data, and there's no way it could have worked well on raw accelerometer samples. In the context of TinyML, where resources are extremely constrained, it is not a given that our board can run a network that does it all by itself (feature extraction + classification), so it is our job to carefully evaluate which approach to follow, given our specific use case.

That said, many optimizations have been adopted to make it possible to run models larger and larger on embedded hardware, so we often can choose. Moreover, deep learning makes it possible to run sophisticated tasks on our cheap, energy-efficient hardware that would otherwise be impossible using only manual features.

Next, let's examine two types of networks that fall under the deep learning umbrella: recurrent and convolutional neural networks.

Recurrent Neural Networks and Long Short-Term Memory

Recurrent neural networks (RNNs) have specific topologies that work best with sequence data. They've been successfully applied to language translation, speech synthesis, video analysis, and, what matters most to us, time series classification.

What makes them different from a "regular" network (e.g., fully connected) is that the hidden neurons have *memory*. That means that when a new input arrives, the neuron uses its internal state to compute the output and then updates that state based on the input and the output (*in addition* to the traditional weights of fully connected and convolutional networks!).

Long short-term memory (LSTM) cells are the most widely used type of neurons in an RNN because of their excellent performances. They were introduced to solve technical problems during the learning steps that prevented the network from achieving satisfactory results over long-duration inputs (vanishing and exploding gradients). Without going too much into the implementation details, here's a quick overview of how these neurons work.

1. Each LSTM cell has an internal state (like the weights in a perceptron) and a *memory state* (another set of weights).

2. When a new input is received, it computes a *forget vector* by combining the input vector and the internal state. Values close to 0 indicate that the corresponding element in the memory state should be forgotten; values close to 1 mean that the element should be retained.

3. It also computes a *candidate vector*, which is how to update the memory state based on the new input. The new memory state is computed as a weighted sum of the old state, the forget vector, and the candidate vector.

4. Eventually, the memory state and the input are used to update the internal state, which produces the output after passing the result through an activation function.

Figure 8-8 is a schematic view of the internal state of an LSTM cell.

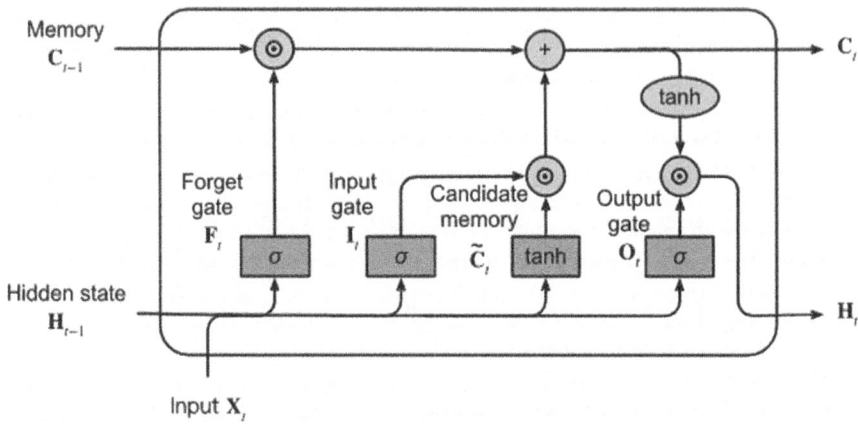

Figure 8-8. LSTM internal structure

How to Train a Recurring LSTM Neural Network

Thankfully, this process is completely transparent to the end user and handled behind the scenes. Creating an RNN network with LSTM cells is achieved with Listing 8-3. Let's train the network on the continuous motion dataset collected in Chapter 4. The accuracy trend on the validation set is plotted in Figure 8-9.

Listing 8-3. Train an LSTM Recurring Neural Network

```
from tinyml4all.tensorflow import RNN
from tinyml4all.tensorflow.layers import LSTM, Perceptron

from tinyml4all.time.continuous.classification import TimeSeries

ts = TimeSeries.read_csv_folder("Chapter4/motion")

ts.label_from_source()
# convert TimeSeries to X and Y training data for NN
```

```
X, Y = ts.as_windows(duration="1s", shift="250ms")

rnn = RNN()
# two LSTM layers with 12 neurons + one Fully connected
rnn.add(LSTM(12))
rnn.add(LSTM(12))
rnn.add(Perceptron(32))

# display network architecture
print(rnn.compile(X, Y, task="classification"))
```

Model: "sequential"

Layer (type)	Output Shape	Param#
lstm (LSTM)	(None, 125, 8)	480
lstm_1 (LSTM)	(None, 125, 16)	1600
flatten (Flatten)	(None, 2000)	0
dense (Dense)	(None, 32)	64032
dense_1 (Dense)	(None, 4)	132

Total params: 66244 (258.77 KB)
Trainable params: 66244 (258.77 KB)
Non-trainable params: 0 (0.00 Byte)

```
# train neural network and display accuracy plot
rnn.fit(X, Y, epochs=50, plot=True)

# print accuracy on the validation set
print(rnn.classification_report())
```

CHAPTER 8   TENSORFLOW FROM SCRATCH

```
              precision    recall  f1-score   support

        idle       1.00      1.00      1.00        52
       shake       1.00      1.00      1.00        91
       slide       1.00      0.86      0.92        92
        wave       0.88      1.00      0.93        91

    accuracy                           0.96       326
   macro avg       0.97      0.96      0.96       326
weighted avg       0.97      0.96      0.96       326
```

```
+-------------------+------+-------+-------+------+
| True vs Predicted | idle | shake | slide | wave |
+-------------------+------+-------+-------+------+
|              idle |  52  |   0   |   0   |  0   |
|             shake |   0  |  91   |   0   |  0   |
|             slide |   0  |   0   |  79   | 13   |
|              wave |   0  |   0   |   0   | 91   |
+-------------------+------+-------+-------+------+
```

```python
# export to Arduino-compatible C++
rnn.convert_to("c++", class_name="LSTM",
save_to="MotionLSTM.h")
```

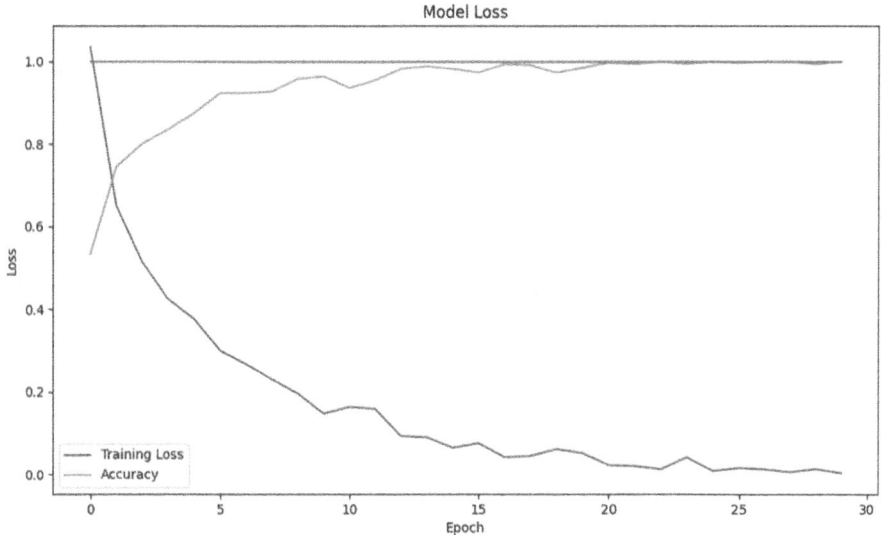

Figure 8-9. LSTM train loss and accuracy

How to Deploy an RNN

Running an RNN on Arduino is the same as running an MLP. You don't have to change a line. The part that would differ is how you create the input vector. A time series is usually built over time by queuing new samples as they are read and discarding old ones. You can handle this manually if you prefer, but the generated code can handle this for you out of the box.

Listing 8-4 showcases how to run an exported RNN to classify the accelerometer data from an Arduino Nano BLE Sense built-in IMU.

Listing 8-4. Run LSTM Network on Accelerometer Data

```
/**
 * Listing 8-4: Classify continuous motion using LSTM model
 *
 * Required hardware: Arduino Nano BLE Sense
 */
```

CHAPTER 8   TENSORFLOW FROM SCRATCH

```
#include "./MotionLSTM.h"
#include <Arduino_LSM9DS1.h>
#include <tinyml4all.h>

// ARENA is the amount of memory to reserve for the model
// larger models needs more memory, but there's not a
// formula to calculate the optimal value
// is a trial-and-error process
#define ARENA 20000

tinyml4all::LSTM<ARENA> lstm;
tinyml4all::LSM9DS1 imu;

void setup() {
    Serial.begin(115200);
    while (!Serial);
    Serial.println("TensorFlow LSTM demo");

    imu.begin();
    lstm.begin();

}


void loop() {

    // read accelerometer and gyroscope
    imu.readAcceleration();
    imu.readGyroscope();

    // append readings to internal RNN queue
    lstm.append(imu.ax, imu.ay, imu.az, imu.gx, imu.gy,
    imu.gz);

    // await until queue is full
    if (!lstm.isReady())
```

```
        return;

    // run classification
    if (!lstm.predict()) {
        Serial.println(lstm.error());
        return;
    }

    // lstm.label holds the name of the predicted class
    // lstm.idx holds the numeric id of the predicted class
    // lstm.value holds the numeric output (for regression
       tasks only)

    // lstm.confidence is the probability of the prediction
    // (from 0 to 1)
    // lstm.confidences is an array with the confidences of
       all the classes
    // (one for each class for classification)

    // lstm.outputsAsString returns each class' score, from
    // 0 to 1 (classification only)
    Serial.print("Predicted class ");
    Serial.print(lstm.label);
    Serial.print(" with confidence ");
    Serial.println(lstm.confidence);
    Serial.print(" > Scores: ");
    Serial.println(lstm.outputsAsString());
    delay(1000);
}
```

If you want to handle the input array by yourself, call lstm.predict(inputs) instead, where *inputs* is the array of input values arranged as At=1, Bt=1, Ct=1, At=2, Bt=2, Ct=2, ... where A, B, and C are the sensors' readings, and t is the time step.

1D Convolutional Neural Networks

Recurring neural networks are not the only architecture that works well with sequential data. 1D convolution neural networks (1D CNNs) are a (usually) faster, leaner alternative to them. This type of network doesn't have memory, as the LSTM cells do, so they may not excel in modeling long-term relationships in the input data. Nevertheless, this is often unnecessary since the strongest connections are between nearby neighbors.

A *1D convolution* is a mathematical operation that takes an input sequence (a list of numbers) and a smaller set of numbers (weights) called a *filter* or *kernel* and then combines them to produce a new sequence. Figure 8-10 illustrates the process.

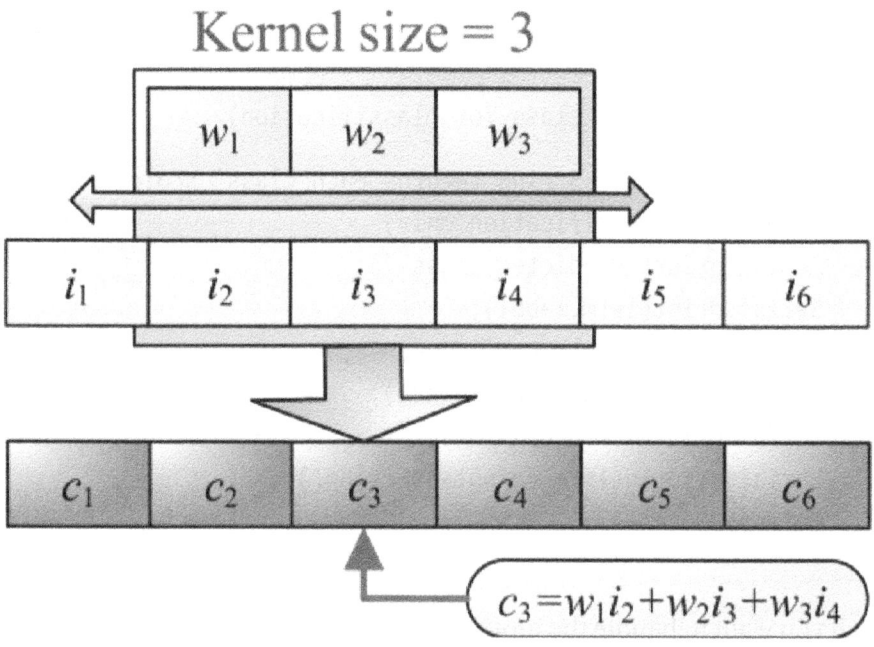

Figure 8-10. *1D convolution*

CHAPTER 8 TENSORFLOW FROM SCRATCH

With reference to the scenario depicted in Figure 8-10, let's begin by considering the first three elements of the input array (in yellow). Each element is multiplied by the corresponding kernel element (in green), and the results are summed up (purple cells). This operation is called the *dot product*.

The kernel then slides to the right by one, and the same operation is repeated, this time with the input elements from 2 to 4. This process loops until the kernel is applied to the last three elements of the input.

The final result is a new sequence that is a weighted aggregation of the input sequence. To compose a 1D CNN, many of these kernels are packed inside a Conv1D layer, and many such layers are stacked one after the other. Between the last Conv1D layer and the output layer, one or more fully connected layers combine the extracted features (see Figure 8-11). The learning process consists of finding the optimal kernel weights that maximize the classification accuracy (in case of classification) or minimize the error (in case of regression).

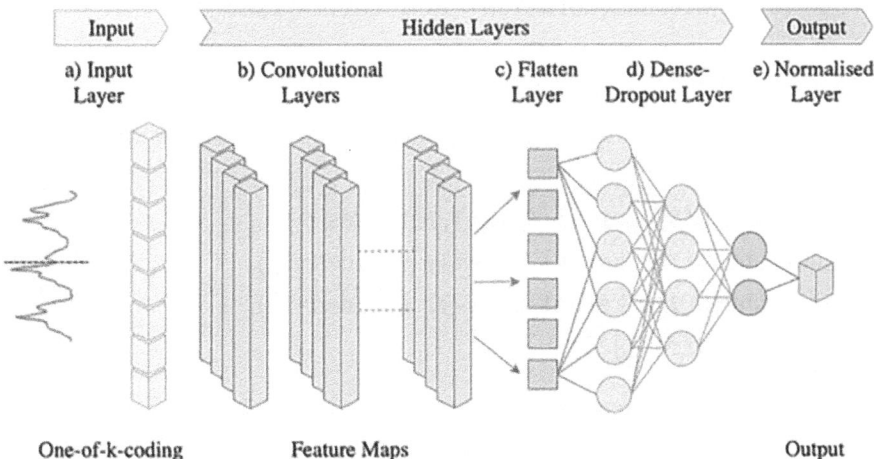

Figure 8-11. Conv1D network topology

CHAPTER 8 TENSORFLOW FROM SCRATCH

How to Train a 1D CNN on Continuous Motion

This section focuses on continuous motion data as a representative example of time series classification. The same procedure works with any data that you can load with TimeSeries.read_csv or TimeSeries.read_csv_folder. See Listing 8-5 for the code and Figure 8-12 for the accuracy plot.

Listing 8-5. Train a Conv1D Network on Continuous Motion Data

```
from tinyml4all.tensorflow import CNN1D
from tinyml4all.tensorflow.layers import Conv1D, Perceptron

from tinyml4all.time.continuous.classification import TimeSeries

ts = TimeSeries.read_csv_folder("Chapter4/motion")
ts.label_from_source()
X, Y = ts.as_windows(duration="1s", shift="250ms")

cnn = CNN1D()

# refer to section "2D Convolutional networks" for what stride is
cnn.add(Conv1D(8, kernel_size=3, strides=2))
cnn.add(Conv1D(16, kernel_size=3, strides=2))
cnn.add(Conv1D(24, kernel_size=3, strides=2))
cnn.add(Perceptron(32))

# display network architecture
print(cnn.compile(X, Y, task="classification"))

Model: "sequential"
```

```
Layer (type)                    Output Shape              Param #
=================================================================
 conv1d (Conv1D)                (None, 109, 8)            152

 conv1d_1 (Conv1D)              (None, 54, 16)            400

 conv1d_2 (Conv1D)              (None, 26, 24)            1176

 flatten (Flatten)              (None, 624)               0

 dense (Dense)                  (None, 32)                20000

 dense_1 (Dense)                (None, 4)                 132

=================================================================
Total params: 21860 (85.39 KB)
Trainable params: 21860 (85.39 KB)
Non-trainable params: 0 (0.00 Byte)
```

```python
# train neural network and display accuracy plot
cnn.fit(X, Y, epochs=50, plot=True)

# print accuracy on the validation set
print(cnn.classification_report())
```

```
              precision    recall  f1-score   support

        idle       0.97      0.98      0.98        60
       shake       0.98      0.97      0.98       104
       slide       1.00      1.00      1.00       106
        wave       0.97      0.97      0.97       105

    accuracy                           0.98       375
   macro avg       0.98      0.98      0.98       375
weighted avg       0.98      0.98      0.98       375
```

CHAPTER 8 TENSORFLOW FROM SCRATCH

```
+-------------------+------+-------+-------+------+
| True vs Predicted | idle | shake | slide | wave |
+-------------------+------+-------+-------+------+
|              idle |  59  |   0   |   0   |  1   |
|             shake |   1  |  101  |   0   |  2   |
|             slide |   0  |   0   |  106  |  0   |
|              wave |   1  |   2   |   0   | 102  |
+-------------------+------+-------+-------+------+
```

```
cnn.convert_to("c++", save_to="CNN1D.h")
```

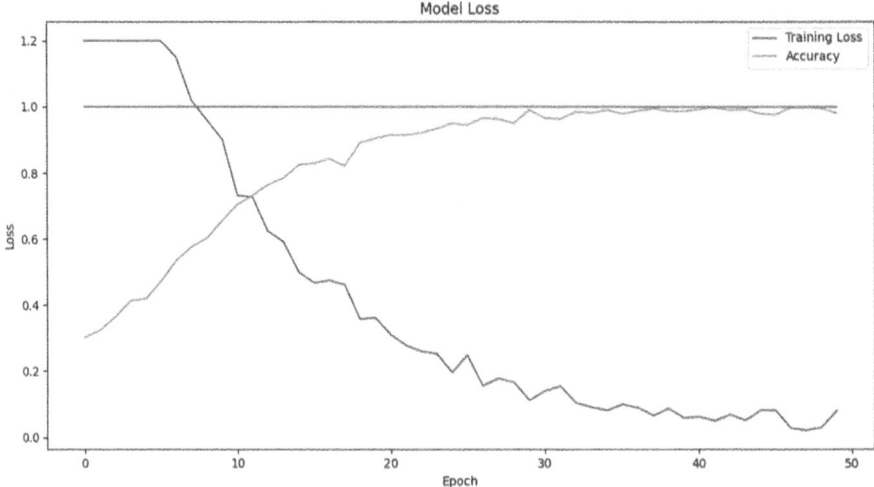

Figure 8-12. *CNN1D loss and accuracy*

How to Deploy a 1D CNN

It turns out that running a 1D CNN in your Arduino sketch is the same as running an LSTM network. Since they both work on time series, everything that applied to it still holds true. And since the generated API is the same, you don't have to change anything—apart from renaming tinyml4all::LSTM to tinyml4all::CNN1D.

2D Convolutional Neural Networks

A 2D convolutional neural network (2D CNN) works on 2D data. 2D data usually means images. Or to widen the application spectrum of CNNs: *image-like* data. By *image-like* data, I mean data arranged in two dimensions with values bounded into a specific domain.

- **MFCC features**: The ones extracted from Edge Impulse in Chapter 6. Pixels represent frequency contribution over time instead of light over space.

- **Electroencephalogram/electromyogram/ accelerometer data**: Every n-dimensional data collected over time can be reshaped into 2D windows (of shape number of features times time steps).

- **Temperature/distance maps**: Thermal cameras (e.g., MLX90640) and grid time-of-flight sensors (e.g., VL53L5CX) produce a 2D output, where each element represents temperature/distance instead of light.

If we abstract the concept of a pixel to encompass every kind of data— not only light—all of this can be rendered as images and thus become a suitable input for 2D CNNs.

The intuition behind the Conv2D operator is that 2D data is *spatially* correlated, and each pixel shares a relationship with its neighbors (called *local receptive field*). Many times, this field is a 3×3 or 5×5 grid. This relationship is encoded in the Conv2D operator weights (called a *kernel*) and is learned from the input dataset. The network can learn many different patterns (e.g., edges, corners, textures) by stacking many kernels together. With respect to fully connected networks, the reduction in the number of weights is drastic and, more importantly, is independent of the number of inputs/outputs. A Conv2D operator of shape 3×3 always stores nine weights, no matter if it applies to 96×96 or 512×512 images.

Downsampling and Stride

Conv2D layers are usually put at the beginning of a CNN. Their job is to extract features from the images that later perceptrons aggregate to produce the network's output. As stated in the relevant paragraph, perceptrons' number of weights grows linearly with the number of inputs. Since images are very "dense" types of data, usually made of thousands of pixels, this poses a problem, both in terms of memory and computation time and accuracy.

Let's consider a 128×128 image and a Conv2D layer with 16 kernels. This produces 128 × 128 × 16 = 262,144 outputs. If we stack a fully connected layer with 32 neurons after it, we end up with ~8 million weights. Then we need an output layer, right? With ten classes, we'd need 80 million more weights! Hopefully, you can see how quickly this becomes an intractable problem. To counter this situation, every CNN incorporates a *downscaling* strategy: the deeper the image moves into the network, the smaller it becomes. In many real-world architectures, the output of the last Conv2D layer has been brought down from 128×128 to 16×16. This downscaling brings a couple more benefits.

- **It allows more kernels at deeper layers.** The more you move along the layers, the more complex the extracted features are. If the first layers can extract edges and corners, the last layers usually learn to recognize shapes and simple objects. Having more kernels there means learning many different shapes, which should be highly correlated to the expected output.

- **It mitigates overfitting.** Reducing the number of inputs is an efficient way to prevent overfitting.

- **It offers translation invariance/noise rejection.** By aggregating local features, small variations in the data are smoothed out, and the overall feature extraction is less sensitive to pixel noise or small translations of objects.

There are two ways to achieve downsampling in a CNN: pooling and strides.

Pooling

A pooling layer is stacked right after a Conv2D layer to downsample its outputs. Its work is to aggregate small regions (usually 2×2) of the Conv2D outputs into a single value: this operation effectively halves the output image size. The most used aggregation operator is *max*, but one can also compute the average of each block (see Figure 8-13).

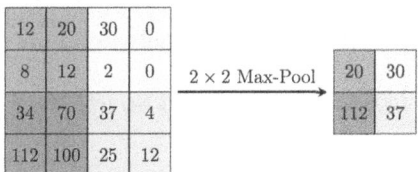

Figure 8-13. Max pooling operator

The reason max pooling is the preferred one in classification is that it emphasizes the most prominent features, often leading to better performance in tasks in which detecting distinct and sharp features (like edges) is crucial.

Strides

The second approach to downscaling is to increase the *stride* of the Conv2D operator itself. The stride indicates how much the kernel moves along the image. By default, it is 1, but this is not mandatory. You could, for example, set the stride equal to the kernel size, in which case the convolution happens on non-overlapping blocks (don't do this; this is just an example!).

Setting the stride to 2 effectively halves the output image size (see Figure 8-14).

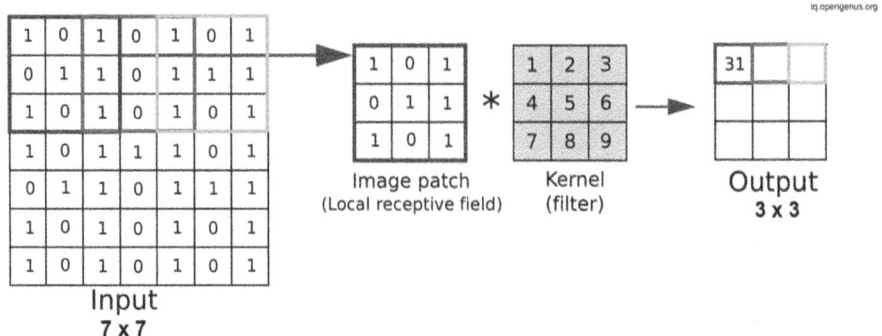

Figure 8-14. How stride works

Table 8-1 summarizes the differences between max pooling and stride 2.

Table 8-1. Comparison of Pooling and Stride

Criterion	Max Pooling	Stride 2 Convolution
Key Benefit	Retains the strongest feature (maximum) Usually performs better than stride	Kernel weights are "aware" of the downsampling during training
Information Retention	Aggressively reduces data (may lose info)	Retains more nuanced information (smoother)
Noise Robustness	More robust to noise (focuses on max)	May be more sensitive to noise or subtle shifts
Computational Cost	Requires convolution with stride = 1, so more computations	1/4 the computations than with stride = 1
Ideal Use Cases	Emphasizing sharp, prominent features	Gradual downsampling while retaining more information

In the context of TinyML, where we must sip every computation and memory allocation, stride has an edge over pooling for the following reasons.

- It requires half the computations along the 2 axis, making it four times faster than full convolution.
- It produces a half-size image straightaway, so less memory is required to store the intermediate results.

How to Train a 2D CNN

Let's train a 2D CNN to classify a toy dataset of images of dogs and cats (Listing 8-6). The dataset is included in the *tinyml4all* package and contains a total of 100 images (50 for each class). Images are 128×128, RGB mode. Figure 8-15 highlights some random images from the dataset.

Figure 8-15. Examples of dogs and cats

CHAPTER 8 TENSORFLOW FROM SCRATCH

> **Note** This is image classification, not object detection! The model won't be able to localize where the dog or the cat is in the image. It only tells if the image depicts one or the other.

Listing 8-6. Train a 2D CNN for Image Classification

```
from tinyml4all.tensorflow import CNN2D
from tinyml4all.tensorflow.layers import Conv2D, MaxPooling2D, Perceptron
from tinyml4all.datasets import Pets

cnn = CNN2D(input_shape=(48, 48))

# example of Conv2 + max pooling
# 8 is the number of kernels
cnn.add(Conv2D(8, kernel_size=3))
cnn.add(MaxPooling2D())

# example of Conv2D with stride
cnn.add(Conv2D(16, kernel_size=3, strides=2))

# stride & max pooling
# (result is ¼ the size)
cnn.add(Conv2D(24, kernel_size=3, strides=2))
cnn.add(MaxPooling2D())

# fully connected layer before output
cnn.add(Perceptron(32))

# display network architecture
print(nn.compile(Pets.X, Pets.Y))

Model: "sequential"
```

```
Layer (type)                  Output Shape            Param #
=================================================================
 conv2d (Conv2D)              (None, 94, 94, 8)       224

 max_pool2d (MaxPooling2D)    (None, 47, 47, 8)       0

 conv2d_1 (Conv2D)            (None, 23, 23, 16)      1168

 conv2d_2 (Conv2D)            (None, 11, 11, 24)      3480

 max_pool2d_1 (MaxPooling2D)  (None, 5, 5, 24)        0

 flatten (Flatten)            (None, 600)             0

 dense (Dense)                (None, 32)              19232

 dense_1 (Dense)              (None, 2)               66

=================================================================
Total params: 24170 (94.41 KB)
Trainable params: 24170 (94.41 KB)
Non-trainable params: 0 (0.00 Byte)
```

```
# train neural network and display accuracy plot
cnn.fit(Pets.X, Pets.Y, epochs=50, plot=True)

# print accuracy on the validation set
print(cnn.classification_report())

              precision    recall  f1-score   support

         dog       0.67      0.73      0.70        11
         cat       0.62      0.56      0.59         9

    accuracy                           0.65        20
   macro avg       0.65      0.64      0.64        20
weighted avg       0.65      0.65      0.65        20
```

CHAPTER 8   TENSORFLOW FROM SCRATCH

```
+--------------------+-----+-----+
| True vs Predicted  | dog | cat |
+--------------------+-----+-----+
|         dog        |  8  |  3  |
|         cat        |  4  |  5  |
+--------------------+-----+-----+
```

```
cnn.convert_to("c++", class_name="CNN2D",
save_to="PetsCNN2D.h")
```

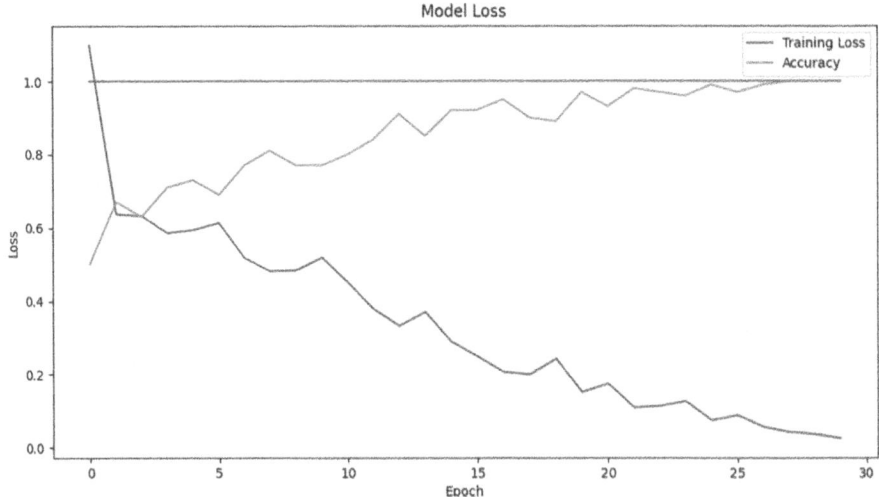

Figure 8-16.   CNN2D loss and accuracy

How to Deploy a 2D CNN

To run a 2D CNN, you need a 2D input. This can be an image, sensor values over time, or multipoint sensor readings. What matters most, though, in this case, is that the TensorFlow runtime *expects 1D input data, even if it represents 2D entities*. For this reason, you must *flatten* your input first and provide it in the form of an *array*, not a matrix. The following are examples of flattening techniques for common data types.

- **Image**: The expected input format for a grayscale image is *row major* (first row's pixels, second row's pixels, etc.). For RGB images, the format is $R_1, G_1, B_1, R_2, G_2, B_2$, and so on. The grayscale case applies to all other image-like inputs.

- **Sensors over time**: The expected input format is *time major* (SensorA$_{t=1}$, SensorB$_{t=1}$, SensorA$_{t=2}$, SensorB$_{t=2}$).

Given these assumptions, Listing 8-7 runs a 2D CNN on a 48×48 grayscale image, stored as a flattened uint8_t array (like the one you could get from an ESP32 or Nicla Vision camera).

Listing 8-7. Run CNN2D on Arduino

```
#include "./PetsCNN2D.h"

// ARENA is the amount of memory to reserve for the model
// larger models needs more memory, but there's not a
// formula to calculate the optimal value
// is a trial-and-error process
#define ARENA 20000

tinyml4all::CNN2D<ARENA> cnn;
uint8_t dog[48*48] = {…};
uint8_t cat[48*48] = {…};

void setup() {
    Serial.begin(115200);
    while (!Serial);
    Serial.println("TensorFlow 2D CNN demo");

    cnn.begin();
}
```

CHAPTER 8   TENSORFLOW FROM SCRATCH

```
void loop() {
    // run classification on dog
    if (!cnn.predict(dog)) {
        Serial.println(nn.error());
        return;
    }

    // .label holds the name of the predicted class
    // .output holds the numeric output (for regression)
    //         or class id (for classification)
    // .outputs in an array with all the outputs
    //         (one for each class for classification)
    // .runtime_ms holds the duration of predictions
    // outputsAsString returns each class' score, from
    //                 0 to 1 (classification only)
    Serial.print("Expected dog, predicted ");
    Serial.print(cnn.label);

    Serial.print(" with confidence ");
    Serial.println(cnn.confidence);
    Serial.print(" > Scores: ");
    Serial.println(cnn.outputsAsString());

    // run classification on cat
    if (!cnn.predict(cat)) {
        Serial.println(cnn.error());
        return;
    }

    Serial.print("Expected cat, predicted ");
    Serial.print(cnn.label);
```
è

```
    Serial.print(" with confidence ");
    Serial.println(cnn.confidence);
    Serial.print(" > Scores: ");
    Serial.println(nn.outputsAsString());
    delay(1000);
}
```

Summary

This chapter was a hands-on introduction to the advanced topic of TensorFlow for Microcontrollers. This framework allows you to unleash the full power of neural networks, but as they say, "With great power comes great responsibility." Tuning a topology that achieves satisfactory accuracy while keeping it as lean as possible for an efficient deployment on embedded hardware can be a hard task.

Neural networks come in many shapes and sizes, and here I tried to showcase the most common configurations and the ones that can be run for sure on Arduino-compatible hardware: fully connected for tabular data, LSTM and CNN1D for time series, CNN2D for images (and similar). Since TinyML is an ever-evolving field, more alternatives should be available to add to your toolset in the near future.

Even if it takes practice, once you gain experience and start to master the art of neural networks, you'll be able to run surprisingly challenging tasks on your board without the need for an external PC connection or cloud services. This supercharges your productivity and redefines what's possible for a tiny device.

APPENDIX A

More Feature Engineering Operators

This appendix introduces a few more feature engineering operators that you can use in your chains when working with tabular and/or time series data and digs more into some of those already introduced in the course of the book. They've been omitted from the main chapters because their use is more sporadic, but they're still valuable and can greatly improve the quality of your model's predictions.

Feature Scaling

This section lists more methods available to perform feature scaling. As a recall, feature scaling is used to alter the domain of the input data, either with a linear transformation (min-max, z-score, robust operators), a non-linear one (Box-Cox and Yeo-Johnson), or an instance-based one (norm). You may want to perform this operation because the machine learning algorithm requires (or works better with) data in the same scale or to make it easier to compare different features that don't share the same scale.

APPENDIX A   MORE FEATURE ENGINEERING OPERATORS

Z-Score Normalization

Also known as *standardization, z-score normalization* is a feature scaling method that consists of rescaling every column so that it has zero mean and unit variance, according to the following z-score normalization formula.

$$x = \frac{x - mean(x)}{std(x)}$$

Being a linear, population-based transformation like the min-max normalization, z-score normalization exhibits the same limits with outliers.

Caution   Z-score normalization is sensitive to outliers!

Listing A-1 shows how to apply this normalization to a table dataset.

Listing A-1. Apply Z-Score Normalization to Table

```
standard = Scale(method="zscore")
table2 = standard(table)
print(table2.describe())
```

	r	g	b
count	150	150	150
mean	0	0	0
std	1,0034	1,0034	1,0034
min	-1,7455	-2,1319	-2,3148
25%	-0,926	-0,6539	-0,4955
50%	-0,0246	-0,1612	0,0243
75%	0,7539	0,5554	0,5441
max	2,5978	7,0049	7,5613

APPENDIX A MORE FEATURE ENGINEERING OPERATORS

Note that the mean is 0, and the standard deviation is (almost) 1.

Robust Normalization

Robust normalization comes into play when your data contains outliers. The word "robust" refers to the fact that it is less sensitive to outliers. Stemming from the z-score normalization formula, this is achieved with a clever substitution, described as follows.

- The mean is replaced by the *median* (the central value after sorting). Outliers are, by definition, at the very ends of the data range, so the central value is not influenced by them.

- The interquartile range (IQR) replaces the standard deviation. This is the range where the central 50% of the data.

$$x = \frac{x - median(x)}{IQR}$$

- Listing A-2 applies robust scaling to a table.

Listing A-2. Apply Robust Normalization to Table

```
robust = Scale(method="robust")
table2 = robust(table)
print(table2.describe())
```

APPENDIX A   MORE FEATURE ENGINEERING OPERATORS

	r	g	b
count	150	150	150
mean	0,0146	0,1333	-0,0233
std	0,5972	0,8297	0,9651
min	-1,0244	-1,6296	-2,25
25%	-0,5366	-0,4074	-0,5
50%	0	0	0
75%	0,4634	0,5926	0,5
max	1,561	5,9259	7,25

Unit Norm

This is an instance-based feature scaling strategy, so it does not compute global statistics. It consists of computing the norm of the sample and dividing each component by it.

$$x = \frac{x}{\|x\|}$$

$\|x\|$ is the norm of x.

Caution   Since the norm is used to scale each component of the sample, it is strongly recommended that, if you also want to perform population-based feature scaling, it be performed beforehand.

There are different orders of norms to choose from.

APPENDIX A   MORE FEATURE ENGINEERING OPERATORS

L1 Norm

L1 norm, also known as the *Manhattan norm*, is calculated as the sum of the absolute values of the component of the vector.

$$\|x\|_1 = \sum_{i=1}^{N} |x_i|$$

The vector is a single sample in our case.

L2 Norm

L2 norm, also known as the *Euclidean norm*, is defined as the square root of the sum of the squares of each component of the vector.

$$\|x\|_2 = \sqrt{\sum_{i=1}^{N} x_i^2}$$

L-max Norm

L-max norm is the maximum of the absolute values of the vector components.

$$\|x\|_{max} = max(|x_i|)$$

Listing A-3 summarizes how to use all the different norms.

Listing A-3. Apply Unit Normalization to Table

```
# L1 norm
L1 = Scale(method="L1")
table2 = L1(table)
print(table2.describe())
```

APPENDIX A MORE FEATURE ENGINEERING OPERATORS

	r	g	b
count	150	150	150
mean	0,4651	0,2874	0,2476
std	0,0774	0,0436	0,04
min	0,3226	0,1951	0,1719
25%	0,375	0,2609	0,2127
50%	0,4944	0,2732	0,2422
75%	0,5312	0,3333	0,2873
max	0,6098	0,3806	0,3188

```
# L2 norm
L2 = Scale(method="L2")
table2 = L2(table)
print(table2.describe())
```

	r	g	b
count	150	150	150
mean	0,0188	0,0124	0,0108
std	0,004	0,0054	0,0047
min	0,0062	0,0055	0,0038
25%	0,0157	0,0084	0,0068
50%	0,0187	0,0102	0,0096
75%	0,0217	0,0165	0,0144
max	0,0332	0,03	0,0253

APPENDIX A   MORE FEATURE ENGINEERING OPERATORS

```
# L-max norm
Lmax = Scale(method="Lmax")
table2 = Lmax(table)
print(table2.describe())
```

	r	g	b
count	150	150	150
mean	0,9981	0,6509	0,5612
std	0,0145	0,2125	0,1852
min	0,8475	0,32	0,3143
25%	1	0,4907	0,4042
50%	1	0,5484	0,4857
75%	1	0,8933	0,7614
max	1	1	0,9167

Box-Cox Power Transform

Box-Cox transform [1] belongs to the family of power transforms [2]. These functions apply non-linear mapping to the input to stabilize variance and make the data more normal distribution-like. The following formula explains how the Box-Cox transform operates while Listing A-4 reports the Python code to apply it.

$$y_i^{(\lambda)} = \begin{cases} \dfrac{y_i^\lambda - 1}{\lambda} & \text{if } \lambda \neq 0 \\ ln(y_i) & \text{otherwise} \end{cases}$$

APPENDIX A   MORE FEATURE ENGINEERING OPERATORS

> **Caution** Since the Box-Cox transform uses the logarithm operator, it only works with strictly positive data!

Listing A-4. Apply Box-Cox Power Transform to Table

```
from tinyml4all.tabular.features import BoxCox

boxcox = BoxCox()
table_2 = boxcox(table)
```

	r	g	b
count	150	150	150
mean	6.4087	2.2936	2.0929
std	1.1713	0.1512	0.1115
min	3.83	1.7369	1.6699
25%	5.3378	2.1957	2.0421
50%	6.5263	2.2897	2.1103
75%	7.3512	2.4006	2.1693
max	8.9006	2.8958	2.5908

Yeo-Johnson Power Transform

Yeo-Johnson transform [3] belongs to the same family as the Box-Cox one. Differently from the former, though, it works with any data, *even negative*. The formula is a bit more articulated, but the concept stays the same: it aims to reduce the input variance (see Listing A-5).

APPENDIX A   MORE FEATURE ENGINEERING OPERATORS

Listing A-5. Apply Yeo-Johnson Power Transform to Table

```
from tinyml4all.tabular.features import YeoJohnson

yeojohnson = YeoJohnson()
table_2 = yeojohnson(table)
```

	r	g	b
count	150	150	150
mean	6.276	2.1292	1.9302
std	1.0779	0.1168	0.0831
min	3.9118	1.7014	1.6159
25%	5.2894	2.0534	1.8922
50%	6.3834	2.1262	1.9432
75%	7.1437	2.2122	1.9873
max	8.5706	2.5908	2.2987

Discretization

Feature discretization is the process of converting a continuous variable into a discrete one. This is done because many machine learning algorithms benefit from this process. The next paragraphs showcase the discretization strategies available in the tinyml4all package.

APPENDIX A   MORE FEATURE ENGINEERING OPERATORS

Binarization

Sometimes, a numeric value brings "too much" information for your use case, and you only need a binary feature.

For example, imagine you're working on a weather station that collects temperature, humidity, and rain. Rain is a continuous variable expressed e.g., in mm/hour. For some tasks, it may be unnecessary to know exactly how much rain fell in a given hour; it may suffice to know whether it rained or not.

This process is called *binarization*, and it requires you to define a *threshold*—values below the threshold are converted to False, and values above are converted to True.

Typically, you don't want to binarize all the columns of your data. In the weather station example, you want to keep temperature and humidity as continuous variables and only binarize the rain. This is why you must specify a column argument in Listing A-6.

Listing A-6. Apply Binarization on the "Rain" Column Only

```
from tinyml4all.tabular.features import Discretize

bin_rain = Discretize (column="rain", threshold=0.1)
table_2 = bin_rain(table)
```

	temperature	humidity	rain
0	22,1	40	False
1	23,3	41	False
2	9,6	60	True
3	9,7	61	True
4	9,5	65	True

APPENDIX A   MORE FEATURE ENGINEERING OPERATORS

Now, the rain column only has True or False values.

If you want to binarize multiple columns, each with a different threshold, you can chain multiple calls, as shown in Listing A-7. This listing also shows that you can put the binarization result into a new column and *flip* the binarization logic (assign True to values *below* the threshold).

Listing A-7. Apply Binarization on Multiple Columns

```
from tinyml4all.tabular.features import Discretize
from tinyml4all.tabular.classification import Chain

# 1. binarize temperatures above 20° and create column "hot"
# 2. binarize humidity above 80% and create column "wet"
# 3. binarize humidity below 20% and create column "dry"
# 4. binarize rain above 0.1

binarize_many = Chain(
    Discretize("temperature", threshold=20, append="hot"),
    Discretize("humidity", threshold=80, append="wet"),
    Discretize("humidity", threshold=20, append="dry",
    flip=True),
    Discretize(column="rain", threshold=0.1)
)
table 2 = binarize_many(table)
```

	temperature	humidity	rain	hot	wet	dry
0	22,1	20	False	True	False	True
1	23,3	30	False	True	False	False
2	9,6	80	True	False	True	False
3	9,7	81	True	False	True	False
4	9,5	85	True	False	True	False

APPENDIX A   MORE FEATURE ENGINEERING OPERATORS

Binning

Binning (also known as *bucketing*) is the process of fitting continuous data into a finite range of possible values ("bins" or "buckets"). The result of this transformation is that small fluctuations in data get smoothed out, and the model becomes less susceptible to overfitting.

In the example of the weather station, we may not need the full resolution of humidity in increments of one. A rough value in increments of ten is probably enough (see Listing A-8).

Listing A-8. Apply Binning to Humidity Column

```
bin = Discretize("humidity", bins=10, append="humidity_bin")
table 2 = bin(table)
```

	humidity	humidity_bin
0	42	4
1	41	4
2	60	6
3	61	6
4	62	6

One-Hot encoding

One-Hot encoding transforms a categorical column (one that can assume only a limited set of values; e.g., the days of the week) into a list of binary values, where only the value in the i-th position is 1 (i represents the position of the value in the list of available ones). This process greatly helps those models that assign an importance score to each feature because it allows different scores to be assigned to different values (see Listing A-9).

APPENDIX A   MORE FEATURE ENGINEERING OPERATORS

Caution   One-hot encoding only works with categorical, discrete data; you cannot encode a real-valued, continuous column.

Listing A-9. Apply One-Hot Encoding

```
from tinyml4all.tabular.features import OneHot

onehot = OneHot(column="fruit")
table_2 = onehot(fruit)
```

	fruit	fruit=orange	fruit=tomato	fruit=zucchini
0	orange	**TRUE**	FALSE	FALSE
1	orange	**TRUE**	FALSE	FALSE
2	tomato	FALSE	**TRUE**	FALSE
3	tomato	FALSE	**TRUE**	FALSE
4	zucchini	FALSE	FALSE	**TRUE**

APPENDIX B

References

Chapter 1

[1] R. Fisher. "Iris," UCI Machine Learning Repository, 1936. [Online]. https://doi.org/10.24432/C56C76.

[2] B. Cabé and B. Cabé, "How I built a connected artificial nose (and how you can too!)," Oct. 4, 2022. https://blog.benjamin-cabe.com/2021/08/03/how-i-built-a-connected-artificial-nose

[3] J. Redmon, S. Divvala, R. Girshick, and A. Farhadi, "You only look once: Unified, Real-Time Object Detection," Jun. 8, 2015. https://arxiv.org/abs/1506.02640

[4] TensorFlow, "TensorFlow," https://www.tensorflow.org/

Chapter 2

None

APPENDIX B   REFERENCES

Chapter 3

[1] A. E. Hoerl and R. W. Kennard, "Ridge regression: Biased estimation for nonorthogonal problems," *Technometrics*, vol. 12, no. 1, pp. 55–67, Feb. 1970, doi: 10.1080/00401706.1970.10488634.

[2] R. Tibshirani, "Regression shrinkage and selection via the lasso," *Journal of the Royal Statistical Society Series B (Statistical Methodology)*, vol. 58, no. 1, pp. 267–288, Jan. 1996, doi: 10.1111/j.2517-6161.1996.tb02080.x.

Chapter 4

[1] "Edge Impulse – the leading edge AI platform." https://edgeimpulse.com/

[2] Wikipedia contributors, "Fourier transform," Wikipedia, Feb. 07, 2025. https://en.wikipedia.org/wiki/Fourier_transform

Chapter 5

None

Chapter 6

[1] "Getting started | Edge Impulse Documentation." https://docs.edgeimpulse.com/docs/pre-built-datasets/keyword-spotting

[2] "Create your Azure Free account or pay as you go." Microsoft Azure. https://azure.microsoft.com/en-us/pricing/purchase-options/azure-account

Chapter 7

[1] M. Sandler, A. Howard, M. Zhu, A. Zhmoginov, and L.-C. Chen, "MobileNetV2: Inverted residuals and linear bottlenecks," Jan. 13, 2018. https://arxiv.org/abs/1801.04381

Chapter 8

[1] TensorFlow, "TensorFlow," https://www.tensorflow.org/

[2] F. Rosenblatt, "The perceptron: A probabilistic model for information storage and organization in the brain.," *Psychological Review*, vol. 65, no. 6, pp. 386–408, Jan. 1958, doi: 10.1037/h0042519.

[3] R. Fisher. "Iris," UCI Machine Learning Repository, 1936. [Online]. https://doi.org/10.24432/C56C76.

[4] J. S. Bridle, "Probabilistic Interpretation of Feedforward Classification Network Outputs, with Relationships to Statistical Pattern Recognition," in *Neurocomputing*, 1990, pp. 227–236. doi: 10.1007/978-3-642-76153-9_28.

APPENDIX B   REFERENCES

Appendix A

[1] G. E. P. Box and D. R. Cox, "An analysis of transformations," *Journal of the Royal Statistical Society Series B (Statistical Methodology)*, vol. 26, no. 2, pp. 211–243, Jul. 1964, doi: 10.1111/j.2517-6161.1964.tb00553.x.

[2] Wikipedia contributors, "Power transform," Wikipedia, Feb. 13, 2025. https://en.wikipedia.org/wiki/Power_transform

[3] In-Kwon Yeo and Richard A. Johnson, "A New Family of Power Transformations to Improve Normality or Symmetry," *Biometrika*, vol. 87, no. 4, pp. 954-959, Dec., 2000, https://www.jstor.org/stable/2673623.

Index

A

Activation function, 163, 263, 268, 275
APDS9960 Sensor, 43, 50, 53
Arduino
    Arduino Sketch, 98–100
    CLI (command-line interface), 196
    FruitChain.h, 98
    library, tinyml4all, 46, 47
    Python to C++, 98
    warnings, 101
Arduino Nano BLE Sense, 43, 134, 196
    board, 50
    built-in IMU, 279
Arduino sketch, 98–100, 196–198
    regression, 124–127
Artificial intelligence (AI), 1, 5, 7, 12, 195, 201
Audio, 38, 39
Audio binary serial output, 198
Audio classification with Edge Impulse, 215–216
Audio data, 196, 210
Audio wake word detection with Edge Impulse
    binary classification task, 194
    microphone, 193
    mics for Arduino boards, 193
    Python script, 199, 200
    tinyml4all Arduino library, 194
Azure dashboard credentials, 203
Azure text-to-speech (TTS), 202–203

B

Backward pass (back-propagation), 260, 261
Bias, 86, 93, 163, 263
Binarization, 308–310
Binary classification, 17, 21–28, 88, 91, 194
Binary classifiers, 18, 19, 90
Binning (bucketing), 310
Box-Cox transform, 305, 306

C

Capture data
    regression, 106–109
    tabular data classification
        collecting data, 48
        quality, 48

INDEX

Capture data (*cont.*)
    RGB components, 48, 49
    save to SD card, 56–60
    serial monitor and manually copy and paste data, 49–53
    serial output from Python, 53–56
  time series classification
    continuous gestures, 136
    copy/pasting from serial, 133
    IMU, 134
    load and inspect the data, 137, 138
    plot time series data, 138, 139
    Python, 173–177
    serial in Python, 135, 136
    serial monitor, 135
Classification chain, 96, 97, 124, 183–185
Classification models
  advantages and disadvantages, 94–97
  decision tree, 84–86
  linearity/non-linearity, 83
  logistic regression, 88–91
  memory requirements, 83
  random forest, 86, 87
  robustness/invariance to feature scaling, 83
  SVM, 91–96
  tinyml4all, 83
  XGBoost, 87, 88
Classification, tabular data
  Arduino, 98–101
  Arduino Nano BLE Sense, 43
  capture data, 47–60
  classification chain, 96, 97
  classification models, 82–96
  feature engineering, 41, 71, 72
  feature scaling, 72–78
  feature selection, 78–82
  load and inspect data, 60–71
  software
    Python virtual environment, 45, 46
    tinyml4all Arduino Library, 46, 47
  TCS3200 color sensor, 44
  workflow, 42
CNNs, *see* Convolutional neural networks (CNNs)
Coefficient of determination, 28
Comma-separated value (CSV) encoding, 49, 51, 52
Computer vision models, 11
Confusion matrix, 24–26, 147, 164, 220, 243
Continuous motion, 145, 157, 165
  classifier, 129, 130
  data, 1D CNN, 284–286
Conv2D operator, 287, 289
Convolution, 217, 290
Convolutional neural networks (CNNs), 214, 216–219, 239, 241
CS, 57
CSV Wizard, 149–152

INDEX

D

Data acquisition, 146, 147, 149, 153, 155
Dataset generation process, 203
Data types
    audio, 38, 39
    images and spatial data, 39, 40
    tabular data, 34–36
    time series data, 36–38
Decision tree, 84–86, 93, 120–122
Deep learning, 11
    disadvantages, 274
    feature extraction, 273
    optimizations, 274
    popularity and accessibility, 273
Dense layer, 158, 162, 163, 268
Deployment, 10, 124, 147, 164–168, 222–225, 245–248
Dimensionality reduction, 15, 33, 67, 211, 214
Downscaling strategy, 288, 289

E

EdgeAI, 5–7
Edge Impulse, 171, 196, 201, 222
    accuracy, 161–163
    audio classification, 215, 216
    audio data and images, 145
    bounding box drawing tool, 238
    classification, 159
    confusion matrix, 220
    CSV Wizard, 149–152
    data acquisition, 235–237
    dataset statistics, 155
    design, 155, 156
    development workflow, 132, 133
    feature extraction configuration, 158
    GUI, 155
    labeling method selection, 237
    labeling queue menu item, 237
    learning block, 158–160
    low code platform, 192
    low-code tool, 145
    spectral analysis block, 157, 158
    synthetic dataset, 209, 210
    time series data block, 156
    training configuration, 218
    train/test split, 154, 155
    upload files, 153, 154
    wake word detection test, 221
    workflow, 145–149
Embedded camera sensors, 230
Episodic motion, 173
Euclidean norm, 303
Exponential operator, 114, 117
Extreme gradient boosting (XGBoost), 87, 88

F

FCNN, *see* Fully connected neural network (FCNN)
Feature bagging, 86
Feature discretization, 307

Feature engineering, 33, 34, 210
   for image classification and object detection, 238, 239
   regression
      monotonic functional mapping, 114–117
      polynomial features expansion, 117, 118
Feature scaling, 71–78, 82, 123, 299, 300
Feature selection
   classification metrics, 78
   RFE, 80–82
   score-based selection, 80
   sequential feature selection, 79, 80
FOMO model, 243
Frequency domain features, 38, 131, 143, 144
FruitChain.h, 98
Fully connected neural network (FCNN)
   characterized, 162
   dense layer, 163
   neurons, 162
   perceptron, 163

G

Generalization, 201, 230
GPUs, *see* Graphics processing units (GPUs)
Gradient boosting, 87
Graphics processing units (GPUs), 3, 12
Ground truth, 16, 119, 161

H

Human activity recognition, 10, 11

I, J

Image generative models, 12
Image recognition, 11, 18
Images and spatial data, 39, 40
Impulse design, 145, 155–156, 239, 240
IMU, *see* Inertial measurement unit (IMU)
IMU sensors, 132, 173
Inertial measurement unit (IMU), 131, 132, 134, 135
Inference/forward pass, 260
Inference time, 85, 86, 90, 225
Instance-based feature scaling strategy, 302
Instance-based scaling, 72
Inter-IC Sound (I2S), 193
Internet of Things (IoT), 1, 9, 35
IoT, *see* Internet of Things (IoT)
I2S, *see* Inter-IC Sound (I2S)

K

Kernel, 217, 241, 282, 283, 287
Kernel trick, 92

INDEX

Keyword spotting (KWS), 11, 191, 201, 209, 213
KWS, *see* Keyword spotting (KWS)

L

Large language models (LLMs), 12, 201
Learning block, 158–160, 215
LIDAR point cloud, 13
LLMs, *see* Large language models (LLMs)
L-max norm, 303–305
Load and inspect data, 206–208
    regression, 109, 110
    tabular data classification
        manipulate table, 63–66
        one file for all classes, 62, 63
        one file per class, 61, 62
        plot data, 66–71
        in Python script, 60
        tinyml4all package, 60
        visual inspection, 60
Local receptive field, 287
Logarithmic compression (dB), 212, 213
Logistic function, 88, 89
Logistic regression, 83, 88–91, 94
L1 and L2 norm, 303
Long short-term memory (LSTM), 274–279
LSTM, *see* Long short-term memory (LSTM)

M

Machine learning (ML), 186
    advantage, 3
    definition, 2
    feature engineering, 33, 34
    ground truth, 16
    inferences, 3
    metrics, 21–28
    microcontrollers (MCUs), 7–10
    overfitting, 29, 30
    regression *vs.* classification, 16–20
    supervised *vs.* unsupervised, 14, 15
    test set, 31, 32
    Tiny (*see* TinyML)
    vs. traditional programming, 2
    training data, 31
    training set, 32
    types of data, 34–40
    underfitting, 30
    validation set, 31, 32
MAE, *see* Mean absolute error (MAE)
Magnetometer data, 174, 175
Make predictions, 94
Manhattan norm, 303
Manipulate table
    apply labels, 65, 66
    DataFrame, 63, 65
    single column, 63
    subset of columns, 64
    subset of rows, 64, 65
Matrix, pair plot, 68–71

INDEX

Max pooling operator, 289
MCUs, *see* Microcontrollers (MCUs)
Mean absolute error (MAE), 27, 28
Mel filter banks, 212
Mel spectrogram, 38, 211–214, 222
Memory constraints, 240, 242
Metrics
    binary classification, 21–28
        accuracy, 21
        precision, 22
        recall, 22
        unbalanced, 23
    multiclass classification, 24–26
    regression, 27, 28
Microcontrollers (MCUs)
    bandwidth, 8
    latency, 8
    low power consumption, 9, 10
    PCs/cloud computing, 7
    privacy, 9
    ubiquitous computing, 9
Min-max normalization, 72–74
ML, *see* Machine learning (ML)
MLP, *see* Multilayer perceptron (MLP)
MobileNetV2, 241, 242
Model accuracy, 161, 164, 219, 220
Multiclass classification, 17
    confusion matrix, 24–26
    Iris flower dataset, 24, 26
    off-diagonal values, 25, 26
Multilayer perceptron (MLP), 255, 261, 263–268

N

Neural networks, 41
    structure, 258, 259
    topology, 258
Noise robustness improvement, 214
Normalization, 72, 83, 85, 300
NVIDIA Jetson Nano, 5

O

Object detection with Edge Impulse
    Arduino-based, 228
    Arduino sketch, 245–248
    deployment, 245
    ESP32-based, 228
    ESP32-S3 camera, 230, 231
    generalization properties, 230
    Model Testing page, 243, 244
    Python side, 232–235
    tinyml4all Arduino library, 229
    width multiplier, 241
OLS, *see* Ordinary least squares (OLS)
1D CNNs, *see* 1D convolution neural networks (1D CNNs)
1D convolution neural networks (1D CNNs), 282, 283
One-hot encoding, 300, 310–311
Ordinary least squares (OLS), 118, 119, 121
Outliers, 27, 66, 74–78, 180, 206, 301
Overfitting, 29–32, 85, 120, 164, 288

P, Q

Pair plot matrix, 68–71
PDM, *see* Pulse-density modulation (PDM)
Perceptron, 163, 261–264, 288
Perceptual alignment, 213
Physical computing, 35
Plot data
   low-quality input, 66
   pair plot matrix, 68–71
   2D scatter plot, 67, 68
   types, 66
Plot time series data, 138, 139
Plotting regression data
   many inputs, many scatters, 112–114
   one input, 111
Pooling layer, 289
Population-based scaling, 72
Proximity meter, 103
Pulse-density modulation (PDM), 193
Python
   serial output from, 53–56
   virtual environment, 45, 46

R

Random forest, 86–88, 93, 120–122
Raspberry Pi, 5, 13, 240
Raspberry Pi Zero, 5, 228, 256
Rate limiting, 182
Reason max pooling, 289
Rectified linear unit (ReLU), 263
Recurrent neural networks (RNNs)
   on Arduino, 279
   topologies, sequence data, 274
Recursive feature elimination (RFE), 80–82
Regression
   Arduino Nano BLE Sense, 105
   capture data, 106–109
   chain, 122, 123
   chain to C++, 124
   characteristics, 121
   coefficient of determination, 28
   color sensor, 105
   decision tree, 120–122
   distance prediction, 103, 104
   distance sensor, 105
   feature engineering, 114–118
   load and inspect data, 109, 110
   MAE, 27, 28
   monotonic functional mapping, 114–117
   OLS, 119
   plot regression data, 110–114
   random forest, 120–122
   ultrasonic sensor, 105, 106
Regression *vs.* classification
   air temperature, 16
   binary classification, 17
   inputs and outputs, 16
   multiclass classification, 17
   multi-label classification, 16
   one-label classification, 16
   one *vs.* all strategy, 18, 19
   one *vs.* one strategy, 19, 20

INDEX

ReLU, *see* Rectified linear unit (ReLU)
RFE, *see* Recursive feature elimination (RFE)
RGB components, 48, 58, 100, 104, 124
RMSE, *see* Root mean squared error (RMSE)
RNNs, *see* Recurrent neural networks (RNNs)
Robust normalization, 301, 302
Robust scaling, 74, 77, 301
Root mean squared error (RMSE), 27, 31, 78
Run LSTM Network on Accelerometer Data, 279, 281
Run MLP on Arduino, 271

S

Score-based selection, 80
SD card, 48, 56–60, 196, 230
Sensor values, 150, 173, 191, 294
Sequential feature selection, 79, 80
Sequential plot of audio samples, 208
Shift parameter, 182
Short-time Fourier transform (STFT), 211–213, 222
Single time series, 179, 181
Spectral analysis block, 157, 158
Spectral features, 146, 157, 158, 164
Spectrogram, 212
STFT, *see* Short-time Fourier transform (STFT)
Strides, 288–291
Supervised *vs.* unsupervised machine learning, 14, 15
Support vector machines (SVM), 91–96
Support vectors, 92, 93
SVM, *see* Support vector machines (SVM)
Synthetic Unknown Words, 205, 206
Synthetic wake word generation, 202, 203, 205, 206

T

Tabular data
  characteristics, 35
  columns represent features, 35
  feature/variable, 34
  fixed schema, 35
  homogeneous data, 35
  regression (*see* Regression)
  row isolation, 35
  rows represent samples, 35
  structured data, 35
T-Distributed Stochastic Neighbor Embedding (t-SNE), 67
Temperature/distance maps, 287
TensorFlow models, 41
  Arduino IDE, search, 257
  deployment ratio, 270–272
  hardware requirement, 256

INDEX

for Microcontrollers, 297
tensorflow-runtime-universal
    Arduino library, 257
validation ratio, 269, 270
Text-to-speech, 195, 202–203, 205
Third-party datasets, 201
Time constraints, 242
Time-domain features, 142, 143,
    168, 171
Time-frequency
    representation, 213
Time-of-flight principle, 106
Time series classification, 273
    capture data, 133–139
    continuous motion classifier,
        129, 130
    deployment
        Arduino library, 164, 165
        cache invalidation, 169, 170
        compilation times, 168, 169
        inference serial output, 168
        run impulse on
            accelerometer data,
            166, 167
    Edge Impulse (*see* Edge
        Impulse)
    feature engineering, 139–144
    IMU, 131, 132
    magnetometer, 132
    no-code tool, 131
    Python
        to C++, 186
        capture data, 173–177
        data labeling, 177
        deploy to Arduino, 186–189
        episodic time series
            classification
            chain, 183–186
        feature engineering, 179–183
        gestures, 172
        IMU sensors, 173
        magnetometer, 172
        media control device, 172
        move next/back, 172
        play/pause, 172
        raise/lower volume, 172
    statistics, 129
    testing, 164
Time series data, 33, 36–38, 129,
    137, 150
Time series data block, 156–157
Time series feature engineering
    advantages and
        disadvantages, 142
    FIFO (first in, first out) data
        structure, 140
    frequency, 141
    frequency domain features,
        143, 144
    machine learning model, 140
    Python
        autocorrelation, 181
        multidimensional data, 179
        one *vs.* rest, 183
        shape metrics, 181
        statistical moments, 180, 181

INDEX

Time series feature
    engineering (*cont.*)
  windowing, 182
  time analyzes the data, 141
  time-domain features, 142, 143
  window logic, 140
  windows of data, 139
TinyML
  definition, 5
  EdgeAI, 5–7
  enabling factor, 4
  federated learning, 4
  hardware, 6
  human activity
    recognition, 10, 11
  image classification and object
    detection, 11
  image generative models, 12
  keyword spotting, 11
  LIDAR point cloud, 13
  LLMs, 12
  on-device learning, 4
  optimization, 4
  predictive maintenance, 12
  quality requirements, 4
  Raspberry Pi Zero, 5
  reliably and satisfactorily, 10
  resource-heavy hardware, 4
  TensorFlow, 5
  tinyml4all, 46, 72, 83, 177
tinyml4all Arduino Library, 46, 47
tinyml4all.tabular.
    classification, 110

2D convolutional neural network
    (2D CNN), 287, 291–296
2D scatter plot, 67, 68

U

Ubiquitous computing, 9
Ultrasonic sensor, 105, 106
Upload data, 149, 153, 209

V

Validation loss, 161, 162
Virtual environment, Python, 45, 46
Visual debugging, 248–252
Voice assistant devices, 192

W

Wake Word Dataset, 204
Wake word detection output, 224

X

XGBoost, *see* Extreme gradient
    boosting (XGBoost)

Y

Yeo-Johnson transform, 306, 307

Z

Z-score normalization, 300, 301

The manufacturer's authorised representative in the EU is Springer Nature Customer Service Centre GmbH, Europaplatz 3, 69115 Heidelberg, Germany. If you have any concerns regarding our products, please contact ProductSafety@springernature.com

Printed and bound by CPI Group (UK) Ltd, Croydon, CR0 4YY

25/03/2026

02078192-0015